HOW TO DRAW A MAP

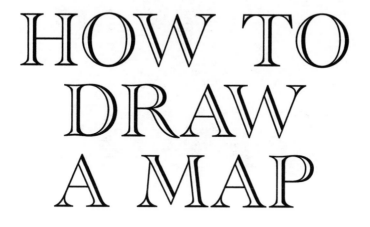

HOW TO DRAW A MAP

Malcolm Swanston
&
Alexander Swanston

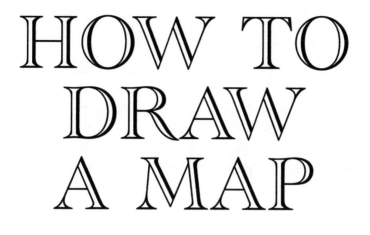

HarperCollins*Publishers*

HarperCollins*Publishers*
1 London Bridge Street
London SE1 9GF

www.harpercollins.co.uk

First published by HarperCollins*Publishers* 2019

10 9 8 7 6 5 4 3 2 1

A catalogue record of this book is available from the British Library.

ISBN 978-0-00-827579-2

Printed and bound in Great Britain by CPI Group (UK) Ltd, Croydon

MIX
Paper from
responsible sources
FSC www.fsc.org **FSC® C007454**

To my wife Heather.

And to Nina, Alexander and Amy.

CONTENTS

CONTENTS

INTRODUCTION

In one way or another, we are all mapmakers at heart. I am sure we have an inbuilt urge to understand the world around us and beyond; how do we, or at least our part of the world, fit into the great scheme of things?

Ever since I can remember, I have been completely absorbed in atlases and maps of all kinds. As a child, I received one particular book that enthralled me for years: *The History of Our Earth*, in which talented artists re-created scenes across double-page spreads of landscapes featuring early dinosaurs, exotic creatures of the desert, human migrations and early civilisations; its well-worn pages are still with me. The moment I first open an atlas, whether it is political, physical, topographical or an historical atlas of any kind on any subject, huge periods of time seem to fly by unaccountably. This affliction has lasted the whole of my lifetime and I appear to have passed it on to one of my children, who is engaged in the production of maps, including the ones used in this book.

My first cartographic undertaking (although I didn't realise it at the time) was not far from my childhood home, at a mile-castle. As the name suggests, these were fortified posts built every Roman mile along Hadrian's Wall. Interspersed between major forts, these milecastles and smaller turrets contained about 50 men to keep watch, usually out to the north. I would set about measuring the visible remains of 'my' castle and working out how its garrison fitted within its walls. My measuring system was a length of rope cunningly adapted to its new use by having a knot tied every yard, and thus I was able to create a five-yard system. Well equipped with a quarter-inch grid in my school notebook, two pencils and a pencil sharpener, and lavishly provisioned with cheese and pickle sandwiches and a bottle of Coke, I set off. I should say *we* set off, as I had recruited my next-door neighbour and school friend Rob to help (facing the unknown alone was too daunting) – a latter-day Mason and Dixon (see page 214).

We measured the milecastle at Gilsland, which, in the 1950s, was in the county of Cumberland but is now part of Cumbria; it is still in my memory six decades later. The milecastle is known to modern archaeology as Milecastle 49 at Harrows Scar. It proved to be 19.8 metres (approximately 30 yards) east to west and 22.9 metres (approximately 32 yards) north to south. It is now in the care of English Heritage. Its monument number, if you are interested in such things, is 13987.

Since the River Irthing was not too far away and stocked with trout prepared to present themselves for capture and the frying pan, this was our next port of call and, I suspect, Rob's real reason for coming along. We succeeded; we came away with the fish and having recorded the edge of the Roman Empire.

My career since then has been almost completely focused on thematic maps – those which display a particular theme, topic or subject of discourse, rather than road maps, with which we are all familiar. Over the course of my life, I have accrued over

50 years of map creation and during these decades cartography has seen the most profound changes, the greatest of which has been technology: the move from physical craftsmanship to the use of computers. Trading the pen, scalpel and scribing tool for software and the mouse marked a significant evolution in the craft of mapmaking.

It seems like a lifetime ago that I received my first fee-paying commission while working for Rolls-Royce Limited. In those far-off days, we cranked up an ancient machine that rejoiced in the name of the 'Illustromat'. It was originally intended as an aid to technical illustration, used to convert a 2D plan into a 3D angled-view 'isometric' illustration. We had discovered that, from a conventional 2D map, we could create a 3D angled view of any chosen part of the earth's surface.

I was required to create a map of the empire of Alexander the Great, to be featured in a new publication on the famous empire-builder. We taped three maps covering the area of Alexander's empire carefully together and placed them on the flat bed of the Illustromat. The operator switched on the machine, we selected the angle of the view required, lights flashed and the operator began to trace the coastlines, rivers and details of the area of the empire. As he manipulated the complex settings, veins stood out on his forehead. He began to sweat profusely and I fled the room to return with hot, sweet tea after the initial drawing was completed. Over this drawing were placed several translucent overlays, and on each overlay a different layer of information was created – coastlines, rivers, lakes, type, arrows and symbols. Slowly, a total image of the map was formed. In this particular case, an airbrush base was created showing the land form, mountains, rolling hills and sweeping deserts.

Map 1. *My first cartographic adventure was to measure a turret on Hadrian's Wall, a tiny piece of this huge structure. I didn't know it then, but I would revisit this subject many times.*

Not long after my first commission, a huge, life-changing project arrived – a request to produce *The Times Atlas of World History*, as its first edition was then known. Part of the delivery from the customer was a set of maps produced by a cartographer called Erwin Raisz. Raisz, born and raised in the Austro-Hungarian Empire, was a civil engineer and architect, who entered the world of cartography after migrating to the

4

United States. Having studied cartography, he went to work for Harvard University, where he was curator of maps for 20 years, during which time he created a significant body of work using a hand-drawn pen-and-ink technique. These maps had a particularly distinctive look – he called them 'physio-graphic' maps – and, because they were hand-drawn, the style was unique to him; a style I found enthralling and inspiring. I also found his tendency to look at a region or country from any direction (by which I mean not always putting 'north' at the top) utterly liberating. He had somehow systematically recreated an updated version of the medieval cartographer's or illustrator's attempts to portray the surface of the planet, sharing their willingness to look at the globe from unconventional directions. Over the last 50 years I have employed some of that methodology to create solutions when explaining thematic historical events.

In *How to Draw a Map*, I aim to introduce you to the craft that has been my life's work. The early part of the book is based on a selection of maps that demonstrate how the world was first explained cartographically. They have been selected from thousands of potential examples and I hope my choice can, in some way, explain this story. Then we move on to maps that are concerned with filling the void – once we know what shape the world is, we will set about peopling the globe and explaining the growth of civilisations and their changing forms.

And finally, this work is dedicated to the contemporary future of mapping and the potential use of artificial intelligence in the permanent updating of cartographic and thematic data.

I am now responsible for, and sometimes guilty of, producing around 100 thematic histories and the largest body of thematic maps ever produced. All this work has been completed with the cooperation of many individuals whose input has been of immense value, and just some of their stories are included in this book.

1

LOST IN EDEN

The instinct to understand the immediate world around us is not exclusive to humans. Animals create food stores and mark with scent; the honey bee directs its buzzing colleagues to a good supply of nectar, carrying a mental map of their home location. It is, however, the human species who became mapmakers.

Early humans developed a process that psychologists call 'cognitive mapping'; the ability to process spatial data from the world around us is part of all of us – well, most of us; there are those who still get lost. Getting lost 40,000 years ago was extremely dangerous. A family or clan well understood the nature of their 'territory' and that of their neighbours, but beyond that was the vast unknown – and entering the unknown 40 millennia ago was for the brave, desperate or foolhardy.

Despite all the problems of making a journey through an unknown world, modern humans, Homo sapiens, demonstrated their adaptive capabilities and began to spread from Africa 125,000 years ago. Around 40,000 years ago, they began to leave visual descriptions of their experience on rock walls or clay, which represented hills, rivers, animals, dwellings and, of

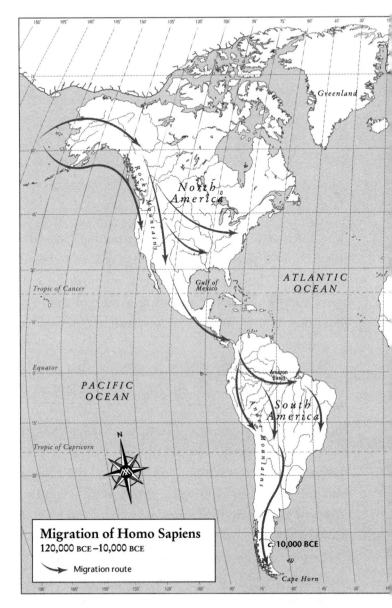

Map 2. *Homo sapiens, modern humans, began leaving Africa 125,000 years ago. Further waves of migration followed 75,000 and 10,000 years ago.*

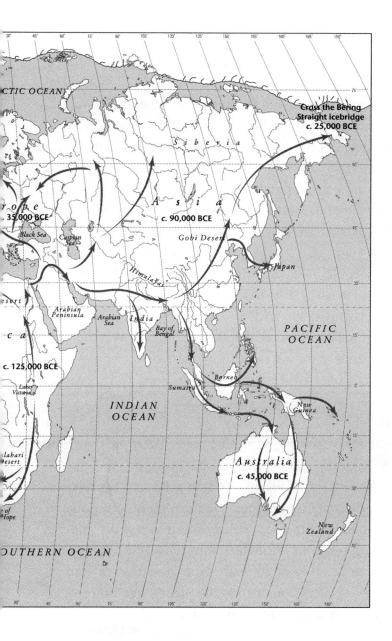

Cross the Bering
Straight icebridge
c. 25,000 BCE

c. 90,000 BCE

35,000 BCE

c. 125,000 BCE

c. 45,000 BCE

CTIC OCEAN

Siberia

Asia

Gobi Desert

Japan

Black Sea

Caspian
Sea

Himalayas

Arabian
Peninsula

Arabian
Sea

India

Bay of
Bengal

PACIFIC
OCEAN

esert

ca

Lake
Victoria

INDIAN
OCEAN

Sumatra

Borneo

New
Guinea

Australia

Kalahari
Desert

of
Hope

New
Zealand

OUTHERN OCEAN

course, themselves. Now, if you were in their shoes, or perhaps moccasins, how would you illustrate your surroundings? Imagine their world and make your marks – what sort of image would you come up with and could it be understood by others? Early man had the same intelligence as today's, though with an utterly different 'world' view. These were the humans who, over thousands of years, migrated and peopled our world. That movement required adaptation to meet the challenges presented by new environments, and there must have been losses because of this movement. Meeting a clan who needed the same resources as your own could, and did, result in warfare, in which losses could be total. Under merciful conditions, the young, fit and healthy survivors might be incorporated into the victors' clan, perhaps with a reduced social status. We are the descendants of these victors and of the survivors.

After first understanding their immediate surroundings, our ancestors also understood the changing seasons and so developed a concept of time, the passing days and, perhaps more importantly, the predictable movements in the night sky. If, in our modern age, you can find a place not polluted by the lights of our civilisation, you might well be overwhelmed by the vastness of your field of vision. The ancients, by experience, will have worked out by the movements of the sun, moon and stars the connection between their position in the sky and the changing of the seasons. And yet, important as it was for the hunter-gatherers to be able to predict the migration of animals and the ripening of various fruits, the emergence of a settled farming lifestyle was perhaps even more critical.

In 1881, working west of Baghdad, then part of the Ottoman Empire, archaeologist Hormuzd Rassam discovered a fragment of a cuneiform clay tablet estimated to be some 4,500 years old.

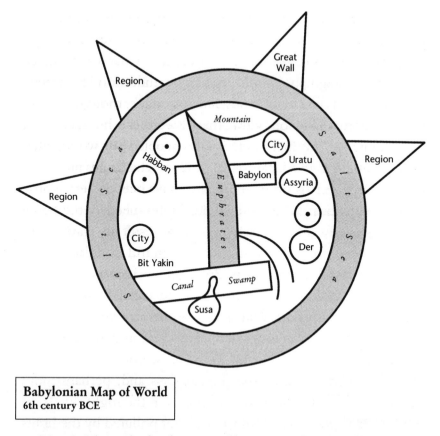

Babylonian Map of World
6th century BCE

Map 3. This is the first known world map, created in the 6th century BCE. It attempts to show the known world as seen from Ancient Babylon.

The tablet was uncovered in what was the ancient city of Sippar, now known as Tell Abu Habbah, in what is now Iraq. It was not until the end of the 19th century that the cuneiform script was translated and the significance of this particular tablet, among tens of thousands, was realised. It is the earliest map we have of the world – not just of a locality or neighbourhood, but of the known world. Looking down on a 'plan' view of the world, the map shows two concentric circles. The space between the circles is called *marratu* or 'salt sea', an ocean that encircles the world. Within the salt sea is the known world and running through the 'world' is a representation that is interpreted as the River Euphrates. The rectangle crossing the river is annotated as Babylon, while there are other symbols labelled as mountains, swamps and canals, and circles to represent important cities and focal points, such as Susa and Urartu.

The Babylonians produced many thousands of such tablets, usually depicting local land ownership – which would have been especially useful when organising land use and implementing taxation. Babylonian mathematicians, some 5,000 years ago, divided a circle into 360 equal segments, or degrees, after they had defined the year as approximately 360 days in length, and that calculation has remained in use into modern times. Those mathematical concepts held real value in the ability to make maps.

2

ANAXIMANDER'S VISION

In 1963, I was introduced to a new world and my vision of our planet took a major leap forward when I embarked on a trip to Greece. While in this historic landscape, with its ruins dating back across the ages, I began to dwell upon the world of the Ancient Greeks. Their concept of the world depended on understanding the natural world that was most familiar to them, one not so different from that known to the Babylonians, except it was a little further west and included a greater familiarity with the Mediterranean Sea and the lands around it.

In the 6th century BCE, in the thriving Ionian Greek city of Miletus, arose a school of pragmatic thinkers who, to some extent, freed themselves from the confines of belief in the Olympian gods; in the words of T. B. Parrington, a British classicist, 'Technology drove mythology from the field.' Reason ruled the day, at least in the cosmopolitan city of Miletus. A citizen of this city, Thales (624–546 BCE), possessed a great intellectual talent and, after making a considerable fortune in the olive oil business, retired and spent his time in the study of the *cosmos*, the Greek word for universe, meaning 'order'. He studied the movements of the sun, moon and celestial bodies,

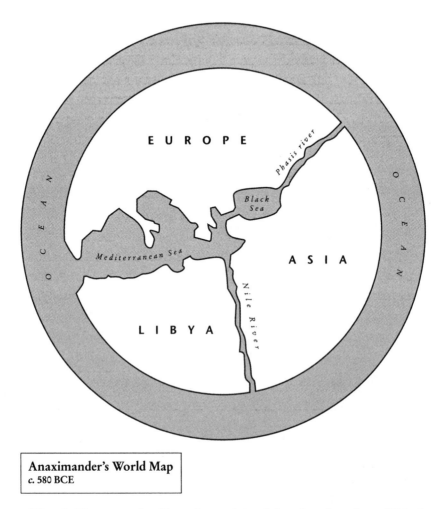

Anaximander's World Map
c. 580 BCE

*Map 4. The geographer Eratosthenes claimed that Anaximander published
the first map of the world. Anaximander was undoubtedly influenced by
earlier maps, including the Babylonian world map.*

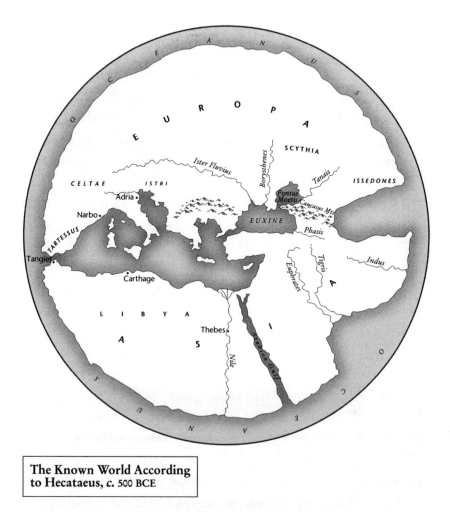

The map labels, reading across the illustration:

OCEANUS

EUROPA

CELTAE · ISTRI · Ister Fluvius · SCYTHIA · Borysthenes · Tanais · ISSEDONES

Adria · Pontus Moetis · Caucasus Mts · Phasis

Narbo · EUXINE

TARTESSUS · Indus

Tangier · Carthage · Euphrates · Tigris · A

LIBYA · Thebes · I · S · A

Nile · Phasis Gulf

OCEANUS

The Known World According to Hecataeus, *c.* 500 BCE

Map 5. Hecataeus is credited with improving upon the work of Anaximander. He created a more detailed interpretation of the world, approximately 80 years after Anaximander.

and also the work of Babylonian and Egyptian researchers. Thales gained notoriety by predicting the solar eclipse of 585 BCE. His view was that by ignoring the mysticism of mythology and observing the natural order, it must be possible to discover the fundamental order of things. He concluded that all things must originate as water and developed the idea of a flat, disc-shaped earth floating on water, the primary substance – not unlike the Babylonian vision of 1,500 years earlier.

A fellow citizen, Anaximander (610–546 BCE), is assumed to have been a pupil of Thales, who was 14 years his senior. Thales – in whose mind observations of earlier societies and Greek philosophy had coalesced – certainly influenced Anaximander. In the theories attributed to him, the diversion from traditional myths and rational thought became obvious – he believed the world floated in the 'infinite' but was a cylinder in shape. Its inhabitable surface, on top of the cylinder, was surrounded by a circular ocean, reminiscent of the Babylonian concept. According to the geographer Eratosthenes, Anaximander was the first to publish a map of the known world, centred on, or at least near, his home city of Miletus. Hecataeus (550–476 BCE), also of Miletus, was inspired by Anaximander's map to create his own version with added detail and produced a prose world history that separated out tradition and myth, relying on factual evidence wherever possible.

ERATOSTHENES AND THE WELL AT SYENE

Alexandria, founded on the northern coast of the Nile delta by Alexander the Great in 332 BCE, prospered after its foundation, far outliving the empire of Alexander, which fell apart after his death. Alexandria came under the rule of the Ptolemies, perhaps the most successful inheritors of his domains. They ruled with a relatively heavy hand, exacting high taxes to pay for a powerful army and navy, but they realised that knowledge was power and they had cultural ambitions. The manifestation of that power was the Great Library at Alexandria. The Ptolemies invested heavily in this fine institution; they even developed a policy that any ship arriving in Alexandria's thriving port, guided into safe anchorage by the great 'Pharos' lighthouse, could be searched for scrolls and learned documents, which would be confiscated from their owners. These would then be carefully copied and the copies would be handed back to their owners, while the originals were placed in the library. Under the patronage of the Ptolemies, Alexandria attracted mathematicians, philosophers and scientists of all kinds. The lights of the Pharos guided valuable trade into Alexandria, but the reach of Alexandria's influence ran well beyond the shores of the Mediterranean Sea – perhaps what modern politicians and diplomats now describe as 'soft power'.

Eratosthenes (275–194 BCE), born in Cyrene and educated in other places including Athens, was a polymath and intellectual philosopher, geographer and mathematician who is reputed to have written a history of comedy, a chronology of major events in the Hellenistic world and also poetry. It was this gifted thinker that Ptolemy III, Euergetes, appointed head of Alexandria's famous library in around 245 BCE. Eratosthenes, at the age of 31, now found himself the librarian in charge of the greatest concentration of knowledge in the known world and was able to sift through new scrolls and reports arriving in the

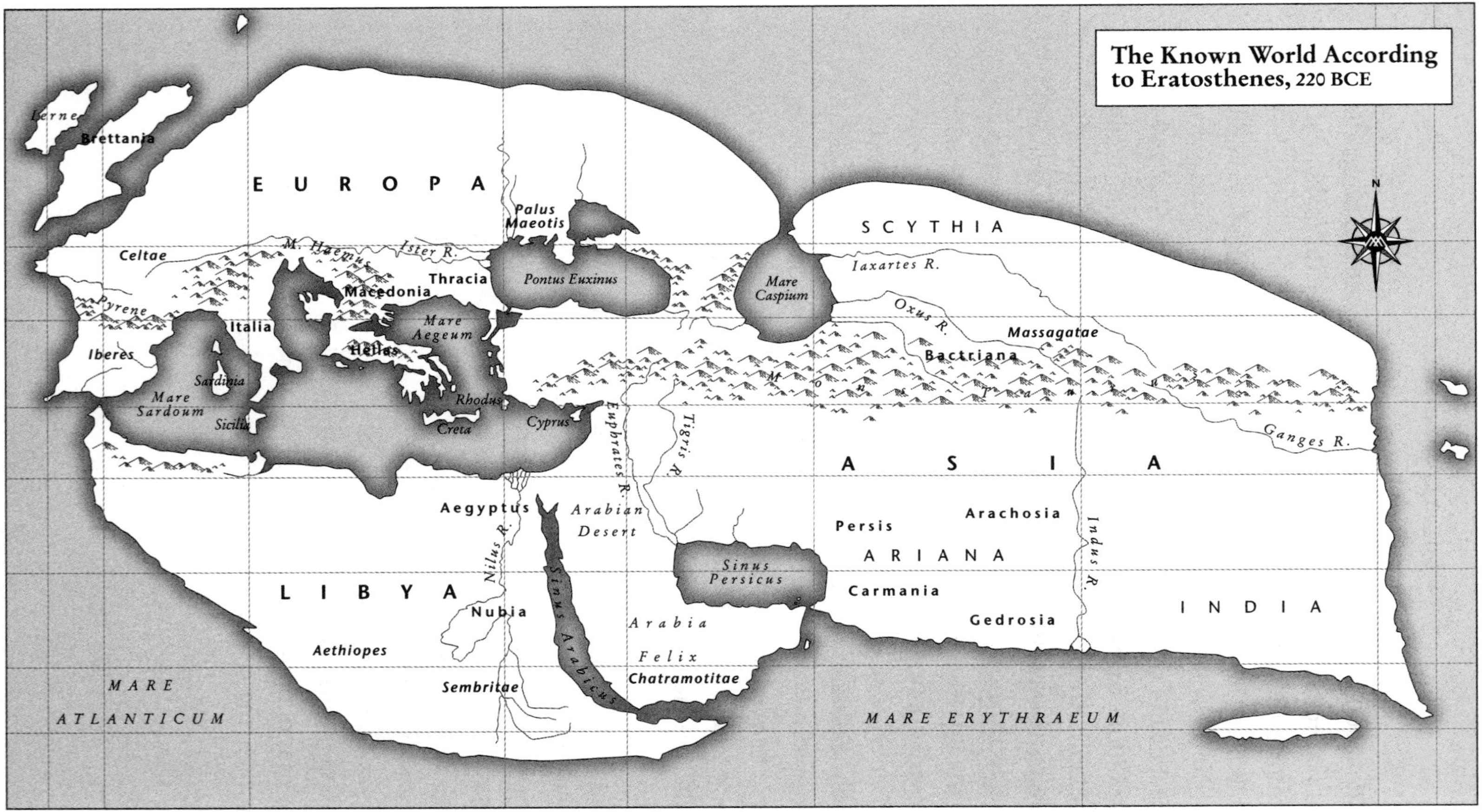
The Known World According to Eratosthenes, 220 BCE
N
Ierne
Brettania
EUROPA
Palus Maeotis
SCYTHIA
Celtae
M. Haemus
Ister R.
Thracia
Pontus Euxinus
Mare Caspium
Iaxartes R.
Pyrene
Macedonia
Mare Aegeum
Oxus R.
Massagatae
Italia
Iberes
Hellas
Bactriana
Sardinia
Mare Sardoum
Sicilia
Rhodus
Creta
Cyprus
Euphrates R.
Tigris R.
Ganges R.
ASIA
Aegyptus
Arabian Desert
Persis
Arachosia
Indus R.
Nilus R.
Sinus Persicus
ARIANA
LIBYA
Sinus Arabicus
Carmania
INDIA
Nubia
Arabia
Gedrosia
Aethiopes
Felix
MARE ATLANTICUM
Sembritae
Chatramotitae
MARE ERYTHRAEUM

Map 6. *Eratosthenes produced a new world map based on his mathematical studies for measuring the earth's circumference. He also divided the world into climatic regions.*

Great Library. This position made Eratosthenes among the most respected scholars in the Greek world.

Among the scrolls was a report of the well of Syene on the Nile River. The report stated that at high noon the sun shone directly down this deep well, lighting its dark waters on the longest day of the year, 21 June. Eratosthenes assumed, based on observations by earlier and contemporary astronomers, that this must mark the northern edge of the tropics. The astronomers of the age believed that the sun, moon and stars rotated around a static earth; from this perspective, they also began to perceive that the sun moved around the earth each day and that, through its annual 365-day cycle, it was higher in the sky in summer than in other seasons of the year. This they called its 'ecliptic'.

Aristarchus of Samos (310–230 BCE) had proposed a theory[*] that the sun was the centre of the known universe and that the earth revolved around it. He had even suspected that the stars were other suns but much further away. However, it is known that Eratosthenes rejected this idea, as Ptolemy did centuries later.

When the ecliptic was plotted, astronomers also noted that the sun was almost always at an angle to the equator, the imaginary line that divided the celestial sphere of the earth. The sun's migrations were calculated to go from 24 degrees south to 24 degrees north of the equator, before returning on the same course; this was called the 'obliquity of the ecliptic'. Now hang in there for a moment …

Though the Greeks spoke of land below the tropics, the consensus of opinion was that people could not live 'down' there. However, the development of the idea of the earth as a sphere

[*] The Greek Heliocentric Theory.

divided by the equator into hemispheres, with lines of latitude marked by the tropics of Cancer and Capricorn, gave geographers their first three reference lines for the making of maps. The northern tropic was named Cancer, which is when the constellation of the Crab (Cancer) appears in the night sky. Likewise, the southern tropic was named Capricorn after the first appearance of the constellation of Capricorn, the horned goat.

Eratosthenes' understanding of the cosmos meant that he could now set about measuring the earth. For his calculations, he considered that the sun's rays would be parallel when they reach earth. Therefore, as the earth was a sphere, the sun's light must hit different parts at different angles. Now he knew that the sun shone straight down the well at Syene and the town's buildings cast no shadow when the sun was almost directly above. At the same moment in Alexandria, the buildings cast a shadow. Therefore, he concluded that if he could measure the angle of the shadow on that particular day at high noon, he might just be able to work out the size of the earth.

Eratosthenes understood that Syene was some 5,000 stades (800 kilometres) due south of Alexandria, and thus on the same meridian (a meridian is half of a circle, from pole to pole that crosses the equator at a right angle). If he could determine the distance between the two locations, this would give him the exact length of the arc of the meridian, the part between the two points, and therefore part of the circumference of the earth. This left him with one unknown in his calculation: if the arc between Syene and Alexandria represented 5,000 stades, what fraction of the full circle of the earth did that represent? Eratosthenes used a gnomon to work this out (a vertical pole used to measure time, set in a public place for everyone to use). When the gnomon's shadow hit the meridian line, the sun was at its zenith. Eratosthenes carefully measured the shadow and the height of the gnomon, which gave him two sides of a triangle. He then drew the third side, and now he could work out the angle of the sun's rays from the top of the gnomon to the edge of the shadow. This turned out to be 7 degrees 12 minutes, almost one-fiftieth of a complete circle, therefore 50 x 5,000 meant the earth was 250,000 stades in circumference, or about 46,000 kilometres. We now know, of course, that the earth is just over 40,000 kilometres around the equator.

Eratosthenes had come pretty close to the correct answer, based on the tools and geographic knowledge available to him. There were mistakes: the earth is not a perfect sphere; Syene is not exactly on the tropic of Cancer (it is 60 kilometres north) and is not on the same meridian (it is 3 degrees 3 minutes east). However, Eratosthenes had established some of the elementary rules in cartography and, for many mapmakers that would follow, the importance of astronomy in calculating your bearings on earth.

During his lifetime, Eratosthenes' researches covered many fields – his colleagues nicknamed him 'Beta' because he covered so much that he always came second (Beta is the second letter of the Greek alphabet). Perhaps there was a hint of jealousy on their part. Other admirers called him Pentathlos after the Olympian athletes who were all-rounders and capable competitors.

In old age, Eratosthenes contracted ophthalmia and became blind, leaving him no longer able to read the scrolls in the Great Library or to observe nature and the heavens. Frustrated, he chose to starve himself to death, dying in 194 BCE aged 82, probably in his beloved library.

3

THE LEGACY OF ROME

The Greeks had made unparalleled progress in the theories of cosmology and geography. The Romans, however, though heavily influenced by Greek theories, were principally concerned with the practical applications of mapmaking. These were adapted and taken into the service of the Roman state and its empire, whether for planning campaigns or travel, administering trade, establishing new colonies or, if possible, superimposing some sort of standard grid so that Rome's expanding possessions could be planned and understood as an interconnected whole – and, of course, exploited (or as we now say, taxed). As the Roman author Strabo said:

> Political philosophy deals chiefly with the rulers, and if geography supplies the needs of those rulers to govern then geography would seem to have some advantage over political science.

Strabo (64 BCE–24 CE) was born in Amaseia, Asia Minor, to wealthy parents who had been movers and shakers in the administration of King Mithridates VI of Pontus. Strabo was

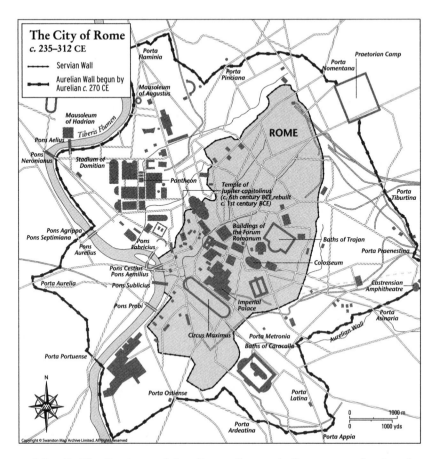

Map 7. *The Servian and Aurelian walls were built to protect the city of Rome, which was becoming increasingly vulnerable to attacks from Germanic tribes in the 3rd century* CE.

Greek by language and learning, and pro-Roman by political inclination. That was probably the right view for a family wanting to hang on to its wealth, since Pontus had recently fallen into the hands of the Roman Republic. In Strabo's lifetime, Rome would extend its control in the region. Strabo was educated in Caria in the city of Nysa under the rhetorician

24

Aristodemus, a teacher who also educated the sons of the Roman general who took over Pontus and, perhaps later, may have known or been influenced by the polymath Posidonius (135–31 BCE). He certainly referred to his work in his later description of Gaul and its Celtic inhabitants.

At about 21 years old, Strabo moved to Rome where he continued his studies under the peripatetic Xenarchus. He studied grammar under the famous Tyrannon of Amisus, also in Pontus, a much-respected authority on geography, and a final influential mentor was Athenodorus Canamotes, a philosopher, originally from near Tarsus, who moved to Rome in 44 BCE where he ingratiated himself with the Roman elite. As well as instructing Strabo, he passed on his contacts in the Roman power structure.

With such an education, contacts and considerable help from the bank of Mum and Dad, Strabo was able to travel and indulge his enquiring mind. He wrote his modestly titled *Historical Notes*, which in fact was a huge undertaking in 43 volumes that, alas, is now lost. However, his *Geographica* ('Geography'), an encyclopaedia of geographical knowledge in 17 volumes, survives – that is, almost survives; part of the end of Volume Seven is lost.

During his time in Rome, Strabo witnessed the end of the Republic, a form of government that had existed for around 480 years. The change dated from 27 BCE when the Roman Senate granted extraordinary power to Gaius Octavius and Marcus Agrippa, great-nephews of the assassinated Julius Caesar (44 BCE). Gaius Octavius adopted the title Augustus and managed to bring peace and stability to what was now the Empire. By now Strabo had returned to Asia Minor, but he was back in Rome by 29 BCE to see the new emperor assume his full power.

25

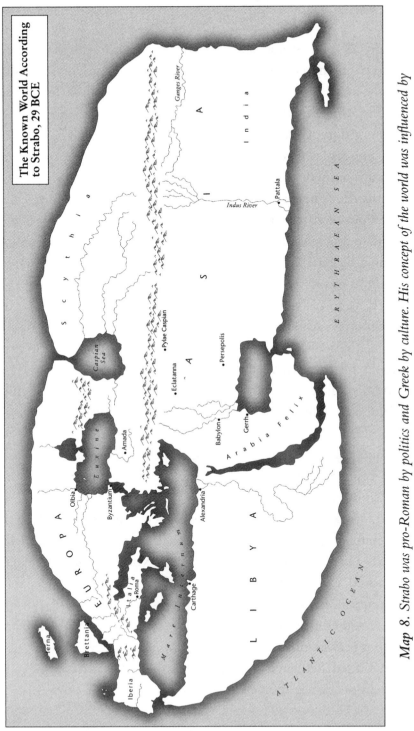

The Known World According to Strabo, 29 BCE

Map 8. Strabo was pro-Roman by politics and Greek by culture. His concept of the world was influenced by Eratosthenes and other Greeks.

In 29 BCE Strabo travelled from Rome to Alexandria in Egypt in the company of his influential friend Aelius Gallus, Prefect of Egypt. Even travelling first class in a well-built galley, this would take around 21 days. Gallus toured his province of Egypt accompanied by Strabo, and part of his mission to the east was to take an expedition to Arabia, an area that was presumed, at least by the Romans, to be full of all kinds of treasures. The purpose of the expedition was to conclude a number of friendship treaties with the local peoples, which of course was intended to benefit Rome. However, the expedition was a failure; after some initial success, there followed a long march to the south through endless deserts, leading to a brief siege of Ma'rib, capital of the Kingdom of Saba. Meanwhile, the accompanying Roman fleet destroyed the port of Aden. All these efforts came to naught after Gallus lost the bulk of his army. Originally 10,000 strong, the weary, sun-scorched survivors retreated to Egypt. The emperor recalled Gallus back to Rome, in some disgrace, but Strabo stayed in Egypt and spent at least some time in the Great Library at Alexandria, the old haunt of Eratosthenes.

Strabo returned to Rome in around 20 BCE, now a Roman citizen, and there settled into writing. He began his surviving work, *Geographica* ('Geography'), the first edition of which was published in 7 CE, far from Rome for some reason. The work was consequently unknown in Rome but widely read in the Romanised east, and maybe that was Strabo's intention, his pro-Roman legacy. After a long gap, there was a second edition in 23 CE, but this was the last year of Strabo's life. He was now back in his home town of Amaseia, and died aged 87, having achieved perhaps the first attempt to assemble the best available geographical knowledge into a single work, which would go on to influence the work of many others.

PTOLEMY

Claudius Ptolemy was probably born in Alexandria around 100 CE. Despite his family name, he was not related to the Egyptian royal family, the founders and protectors of the Great Library. It was in this city, some 400 years after its foundation, that Ptolemy created his masterwork, *Geographite hyphegesis* ('Guide to Geography'), as it was later known among its readership.

The Great Library at Alexandria, dedicated to the Nine Muses of the Arts, had been destroyed some 148 years before Ptolemy's birth during Caesar's Civil War in 48 BCE, when his forces were threatened by the Egyptian navy. During attempts to destroy that threat, Caesar's own ships were burnt and, according to some descriptions, the resulting conflagration spread to the Great Library, destroying much of its invaluable contents.

Ptolemy sat among the patched-up halls of the library compiling his *Geography* in around 150 CE. Much of his writing relied on earlier work, especially that of Marinus of Tyre (70–130 CE). His study was written in Greek – as were most of its precursors – on a papyrus roll over eight sections or 'books', and in it he reviewed the combined total research of the classical world to date. The outcome of this monumental work was to define mapmaking, at least in the West, for the next 2,000 years.

Ptolemy regarded himself firstly as a mathematician, astronomer and philosopher. He did not refer to himself as a geographer. Indeed, the Greek spoken by Ptolemy had no word for geography. What we call a map, he called a *pinax*, or he may have used the phrase *periodos ges*, a 'circuit of the earth'. In time, these terms would be replaced by the Latin *mappa*.

Ptolemy had established his credentials, first in astronomy, writing a work called *Ho megas astronomos* ('The Mathematical Collection'), later known as the *Almagest* (Arab astronomers used the Greek superlative term *Megiste* for this work, and when the definite article in Arabic, '*al*', was prefixed it became *Almagest*).

The *Almagest*'s 13 books deal in detail with astronomical concepts, the stars, the solar system and other observable objects. In them, Ptolemy produced his geocentric theory that placed the earth at the centre of the universe, often known as the Ptolemaic Cosmology, and this view was held by the majority of observers and thinkers until the heliocentric, sun-centred theory developed by Copernicus 1,300 years later.

In the *Almagest* Ptolemy wrote:

I know that I am mortal by nature, and ephemeral; but when I trace at my pleasure the windings to and fro of the heavenly bodies, I no longer touch earth with my feet, I stand in the presence of Zeus himself and take my fill of ambrosia.

He obviously felt some passion for his work. Now we come to Ptolemy's *Geography*. He approached this work by first gathering the learning from the Babylonian, Greek, Roman and Persian worlds and applying this accumulated knowledge into a mathematical framework.

A large part of Book 1 describes how to draw the maps using his 'projection' – his coordinates are in degrees: 360 to complete the circle, 60 minutes to a degree, just as we still use. His east–west longitudes are measured eastwards beginning with 0 degrees at a point just west of the Canary Islands, known to him as the 'Blessed Isles'. The north–south latitudes are measured heading north from the equator, again like our own today. He did, however, mark his degree sign with an asterisk: 10*, not 10° as we now use.

His skill was in devising this mathematical formula that allows the rendering of the earth's sphere onto a flat surface. This was an improvement on past 'world' map projections, given that all maps are some kind of compromise when rendering a sphere onto a flat sheet of papyrus or paper. He also placed north at the top, an orientation once again familiar today. This was the first example of a conical map projection. Ptolemy had succeeded in providing a simple, reliable method of drawing a world map.

The complete text drew on all the works available to Ptolemy in Alexandria, such as Tacitus with his descriptions of Gaul. Ptolemy laid down rules for mapping local features such as cities, harbours and farms. He stressed the importance of astronomy and mathematics to geography, creating a system based on the unchanging features of the sun and the stars, and demanding that longitude and latitude be used to fix locations of geographical features: mountains, estuaries, settlements, etc. By adhering to this system, mapmakers ensured their maps were accurate and could be systematically reproduced. That is, of course, as long as the information supplied was accurate – a problem the modern mapmaker still encounters.

The completed text of *Geography* is more than a list of geographical coordinates. Ptolemy chose to review and carefully analyse the descriptions available to him, then selected the most reliable, but he still left a warning as to the reliability of some descriptions. The mapmaker is only as good as the information he has to hand. This gazetteer was eventually to list some 8,000 locations, according to their latitude and longitude, beginning in the west with the British Isles, moving eastwards across Europe towards Asia Minor and ending with India in the east, his known world. It is not known if Ptolemy produced maps himself to go with his geographic gazetteer; if he did, none have survived.

Ptolemy divided the globe's circumference into 360 degrees (the Babylonian sexagismal system); every degree was measured in units of 60. He estimated each degree at 500 stades (2,700 kilometres), a total of 180,000 stades. He envisaged his world as smaller than Eratosthenes' calculations; however, he did argue that the world's inhabited regions were larger than many believed, reaching from the Fortunate Isles in the west to Cattiagara in the east, believed to be somewhere in modern north Vietnam. From north to south was measured at 40,000 stades from Thule at 63 degrees north to Agisymta in sub-Saharan Africa at 16 degrees south. Ptolemy's world had blurred edges based on theories, stories and guesswork. For instance, he believed that a large, unknown continent existed in the southern hemisphere in order to balance Europe and Asia in the northern hemisphere and that the Indian Ocean was enclosed by Africa, extending eastwards to join up with Asia. This last feature appeared on Ptolemaic maps even after the Portuguese had sailed around Africa into the Indian Ocean.

After Ptolemy's time, whatever was left of the Great Library of Alexandria was lost through invasion and war, a fact that haunts the enquiring mind to this day. However, some works

The World According to Ptolemy
100 CE

Map 9. Ptolemy's maps were the first to use longitudinal and latitudinal lines as well as specifying terrestrial locations by celestial observations.

survived and *Geography* was one of them. The earliest known copy of *Geography* that exists is in Arabic and dates from the 12th century, preserved by Islamic scholars who reintroduced it into Europe, where it was translated from Arabic into Byzantine Greek in the 13th century and then into Latin in the 15th century. The work made it into print in 1475. The Ulm edition, printed in 1482 in Germany, included woodcut maps of new discoveries like Greenland, which was placed using the Ptolemaic principles of its longitude and latitude. Now, 1,300 years after his death, Ptolemy was a bestseller as his books poured off the printing presses.

Ptolemy's mathematical scientific system made the chaotic world understandable by the application of methodical geometric order. Thousands of descriptions made by sailors, travellers and generals, with all their sense of wonder at the almost endless diversity of landscapes and peoples, came together as an understandable whole. Ptolemy's achievement lasted into and past the Renaissance. I still plot my maps in degrees and minutes. I still check gazetteers of verified locations – now online, of course.

> These things belong to the loftiest and loveliest of intellectual pursuits, namely to exhibit to human understanding through mathematics both the heavens themselves in their physical nature (since they can be seen in their revolution about us), and the nature of the earth through a portrait (since the real earth, being enormous and not surrounding us, cannot be inspected by any one person either as a whole or part by part).
>
> Ptolemy's *Geography*, Book I

The classical world in which this achievement was created, though, was now under threat.

4

THE ROAD TO PARADISE

By the beginning of the 4th century CE, the empire in the West was perhaps less 'Roman': many of its provinces were in the ownership of federated tribes – semi-independent kingdoms concerned with their own politics. The Roman state was a less cohesive entity. In the East, Alexandria, still a centre of learning, though much reduced, also became a centre of revolt. The museum buildings, by now over 500 years old and not in the best of condition, were finally destroyed, though the Great Library building still survived. In 391 CE, the library finally met its end when a Christian mob broke in, burned the almost irreplaceable contents and turned the building into a church, a triumph of faith over reason.

Meanwhile, a few years earlier at the opposite end of the empire along Hadrian's Wall in Britannia, Magnus Maximus, who was commander of Britain, withdrew troops from northern and western Britain in pursuit of his own ambitions for imperial rule, usurping power from Emperor Gratian. His attempts ultimately failed, being defeated by Emperor Theodosius, and Maximus was finally executed in 388. With his death, Britannia came back under the direct rule of Theodosius, that is until 392 when another usurper, Flavius Eugenius, made another bid for imperial power. Again, after just two years his short-lived rule over the West failed when Theodosius marched from Constantinople at the head of his army and defeated Eugenius at the Battle of Frigidus in September 394. Eugenius was captured and executed as a criminal. The following year, 395, the victorious Theodosius died, leaving his 10-year-old son Honorus as Emperor in the West. However, the real power in the West was in the hands of Flavius Stilicho, a highly experienced general who had risen through the ranks. In 402 it was his decision to finally strip Hadrian's Wall of its remaining garrison, and possibly other troops in Britain, to face wars with the Ostrogoths and Visigoths on the Continent.

Meanwhile, the Romano-Britons now dispensed with imperial authority. In 407, they selected Flavius Claudius Constantinus, or Constantine III, as their leader, who now declared himself the Western Roman Emperor, gathered the last Roman troops in Britain and headed for Gaul. Sixty-six years later, in the West, Rome was gone, replaced by a collection of 'Barbarian' kingdoms. The new kingdoms lived among the remains of a once great empire. The skills needed to repair and maintain roads, bridges, aqueducts and great buildings were lost. The Anglo-Saxon poem *The Ruin*, written by an unknown author, looks upon the moss-covered buildings with a sense of wonder:

> *Wondrous is this wall-stead, wasted by fate.*
> *Battlements broken, giant's work shattered.*
> *Roofs are in ruin, towers destroyed,*
> *broken the barred gate, rime on the plaster,*

walls gape, torn up, destroyed,
consumed by age, Earth-grip holds
the proud builders, departed, long lost,
and the hard grasp of the grave, until a hundred generations
of people have passed. Often this wall outlasted,

hoary with lichen, red-stained, withstanding the storm,
one region after another; the high arch has now fallen.

The wall-stone still stands, hacked by weapons,
by grim-ground files.

Along with Roman infrastructure, the written word, the scrolls of study, were largely lost, at least in the West. As the natural science of the Greeks and Romans faded and almost disappeared, some 'stories' written in the 3rd century survived and became part of the early medieval world view. One such work was the product of Caius Julius Solinus, whose speciality was the study of grammar. His book *A Collection of Memorable Facts* comprises 1,100 descriptions that range from direct biblical descriptions to tall tales of Africa where the shadows of hyenas robbed dogs of their ability to bark. It was a great collection of myths – exaggerated travellers' tales borrowed from many previous authors, with just enough geographical reality to give this masterpiece of disinformation a kind of believable life of its own. In the 6th century it was revised and republished under the title *Polyhistor*, meaning 'many stories'.

The Christian faith spread around the Mediterranean from its place of origin in Palestine. It was a religion that, in a changing world occupied by usurpers, barbarian invaders, plagues and famine, offered at least the promise of a better life in the next world – the afterlife. The prevailing Greco-Roman religion did not offer any of that – it demanded sacrifice. The cults offered no guidance for the living of a good life; the underworld was not a place of peaceful eternity.

The Christian world still had a place for the Devil and leagues of demons. By the Edict of Thessalonica, 380 CE, Christianity was confirmed as the religion of the Roman Empire. Now, with official sanction, the new faith began its work: the suppression of pagan beliefs. Unfortunately, most of Greco-Roman scientific research was included in the works that were destroyed or suppressed. The Academy School of Philosophy, which had roots going back to 387 BCE, finally closed its doors in 528 CE, after 916 years of considered thought (though with interruptions), its teachers chased away, hunted down as pagans. The new religion destroyed far more than it saved, but in all this chaos, a new world view evolved that was based on faith. This would change the mapmakers' approach to representing the world – at least for a few hundred years.

Sebastian Munster (1448–1552) was a theologian and Hebrew scholar who taught at the University of Heidelberg. Like many learned people of his age, he also had an interest in geography and mapmaking. He produced two major works: the first, in 1540, was Ptolemy's *Geography* with 48 woodcut maps. He carved the names of places, cities and states on removable blocks, which enabled the map to be changed and updated without recarving the entire map. His next cartographic work was *Cosmography*, published in 1544, in which he mapped each continent separately and listed the sources upon which his maps were compiled. Moreover, the blank spaces were 'decorated' with strange fictional races and creatures that featured on the maps of Solinus. These went on to be copied onto maps produced up until the 18th century.

The disinformation spread by the works of Solinus had already been compounded by 'Christian geography'; one such work was written in the 500s by Cosmas Indicopleustes, who believed that all things – the nature of the universe, all living things and the form of the earth – could be found in the Holy Scriptures.

In his earlier life, Cosmas had been a successful merchant trading over much of the known world, around the cities of the Mediterranean and as far east as Ceylon (Sri Lanka). He converted to Christianity and eventually settled into the cloistered life of a monk, retiring to a monastery in the Sinai desert. He produced a work called *Christian Topography*, in which he looked only to the scriptures, describing the world as a flat parallelogram, with Jerusalem at its centre. It was written in Ezekiel 5.5: 'I have set Jerusalem in the midst of the nations and countries that are round about her.' Cosmas argued that in the far north of the inhabited world there was a great mountain around which the sun and moon revolved, creating night and day. Beyond the centre lay a great ocean, and beyond that were other lands where, before the biblical flood, people lived but that were now uninhabited and inaccessible. Beyond these empty lands arose the four walls of the sky meeting in the dome of heaven, the ceiling of the tabernacle. In his *Christian Topography*, Cosmas berates arrogant and sceptical scholars who:

attribute to the heavens a spherical figure and a circular motion, and by geometrical method and calculations applied

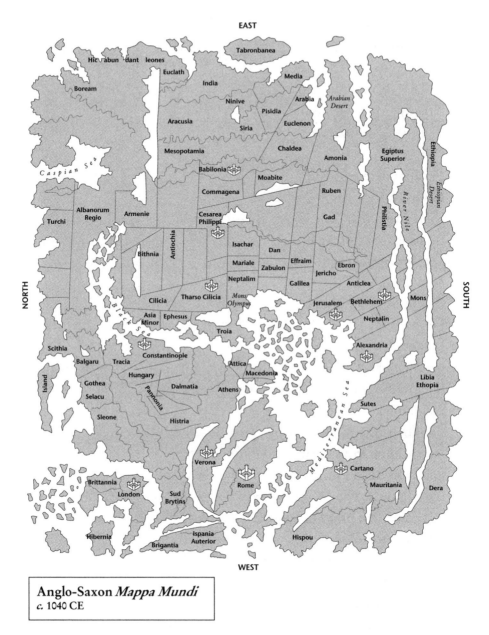

EAST

Tabronbanea

Hic abun dant leones

Euclath

Boream

India

Media

Arabia

Arabian Desert

Ninive

Pisidia

Aracusia

Siria

Euclenon

Mesopotamia

Chaldea

Amonia

Egiptus Superior

Ethiopia

Caspian Sea

Babilonia

Moabite

Commagena

Ruben

Albanorum Regio

Armenie

Cesarea Philippi

Gad

Philistia

River Nile

Ethiopan Desert

Turchi

Bithnia

Antiochia

Isachar

Dan

NORTH

Cilicia

Mariale

Zabulon

Effraim

Jericho

Ebron

Neptalim

Galilea

Anticlea

Tharso Cilicia

Mons Olympus

Jerusalem

Bethlehem

Mons

Asia Minor

Ephesus

Neptalin

Troia

Black Sea

Alexandria

Scithia

Constantinople

Balgaru

Tracia

Attica

Libia Ethopia

Hungary

Macedonia

Island

Gothea

Dalmatia

Athens

Selacu

Pannonia

Sutes

SOUTH

Sleone

Histria

Mediterranean Sea

Verona

Brittannia

Rome

Cartano

London

Sud Brytins

Mauritania

Dera

Hibernia

Ispania Auterior

Hispou

Brigantia

WEST

Anglo-Saxon *Mappa Mundi*
c. 1040 CE

*Map 10. The original of this map was created in around 1040 and
contains the earliest-known vaguely realistic depiction of the British Isles.*

to the heavenly bodies, as well as by the abuse of words and by worldly craft, endeavour to grasp the position and figure of the world by means of the solar and lunar eclipses, leading others into error, while they are in error themselves, in maintaining that such phenomena could not represent themselves if the figure was other than spherical.

Cosmas determinedly went on to discourage any consideration of Greek thought regarding the possibility of people inhabiting the Antipodean side of the spherical earth, for they 'could not be of the race of Adam'. Did not the scriptures refer to the four corners of the earth? The Apostles were commanded to go out into the world and preach the Gospel to every creature; they could not reach the Antipodes and therefore such a place could not exist.

Now let's take a look at the early medieval world as portrayed in Christian maps. I use the word 'Christian' very loosely, to describe a region and time, as in mapmaking cultures tended to overlap and influence each other. About 1,100 *mappa mundi*, or world maps, from this period are known to exist. The *mappa mundi* come in a number of groups, which include zonal maps, a group of diagrammatic maps that divide the world sphere into five climactic zones. Only two of these zones were believed to be inhabited: the northern temperate and the southern temperate, which were separated by an imaginary ocean along the equator.

Zonal maps attempted to fit landforms into the zonal concept, which emphasises the separating equatorial ocean and places north at the top, as we would view a world map today. They looked back to a Greek tradition where human existence was shaped by the natural world, and therefore it was believed that there was an unknown and inaccessible theoretical race occupying the southern temperate zone, which corresponded with the northern temperate zone. For the Christian mind this was an unresolved problem; this race was not mentioned in the Bible, so was it created by God?

Next (though in no particular order) are T-O (sometimes called the 'Tripartite') maps These offered a simplified description of geography, showing only the inhabited regions of the world as known to Roman and medieval scholars. The world map was illustrated within a circle, with the three land areas – Europe, Africa and Asia – divided by a T shape, representing water that reached out to a circular ocean. Most of these maps placed east (*oriens* in Latin) at the top, giving us the term 'orienting'. These maps also resolved the problem of an unknown race by completely ignoring the possibility, at least

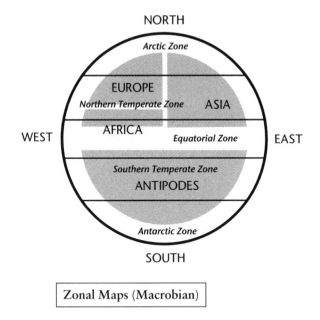

Map 11. *These maps are sometimes called 'Macrobian', after the Roman administrator and author Ambrosius Aurelius Theodosius Macrobius.*

geographically. Isidore, Archbishop of Seville (580–636 CE), used this style of map in his epic *Etymologiae* ('Origins'), a collection of works from antiquity that would probably have been lost without his efforts.

In the maps known as Beatus, or Quadripartite, east is again shown at the top, where the Garden of Eden is located – a focal point of Christian belief. These maps show the three known continents plus a new and unknown continent to the south, hence the name Quadripartite. This fourth continent is often called *Antipodes*. The Beatus maps are named after Beatus of Liébana, a monk and theologian with a keen interest in geology. In the years running up to his birth in around 730 CE the bulk of Visigothic Spain had been overrun by Muslim invaders of the Umayyad dynasty, with the exception of the

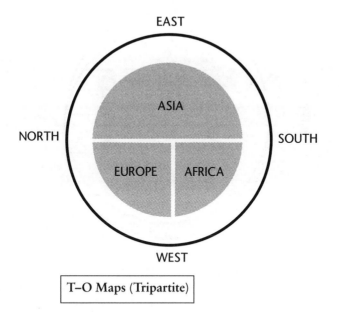

T–O Maps (Tripartite)

Map 12. These maps illustrated only the habitable portion of the world, as it was known in Roman and medieval times.

kingdoms of north-west Spain, the home of Beatus. The most accurate mapmakers of the day were Arabic, and their maps would have influenced the young monk, who is now best remembered for his work *Commentary on the Apocalypse of St John*, published in 776 with revisions in 784 and 786. The Beatus maps that now exist are believed to derive from an original now lost to us.

The Complex, or Great, maps are perhaps the most famous manifestations of the *mappa mundi* world maps. The Ebstorf map, made by Gervase of Ebstorf following the T-0 format in the 12th century, was rediscovered in a convent in Ebstorf in 1843. It was painted on 30 goatskins and showed detailed landforms with illustrations from classical antiquity, as well as

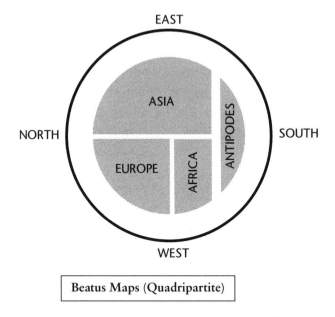

Beatus Maps (Quadripartite)

Map 13. These maps represent a kind of amalgam of Zonal and T–O maps and include a fourth, unknown continent, sometimes labelled 'Antipodes'.

images from biblical history. The head of Christ is shown at the top – the east – of the map. The original was lost in an allied bombing raid in 1943, but a number of copies still survive.

The largest map of this style known to still exist is the Hereford *mappa mundi*, created around 1300. It is drawn and painted on a single sheet of vellum, 158cm by 133cm, and illustrates fifteen biblical events, five scenes from classical mythology, 420 towns and cities, plus people, plants and animals. Again, east is at the top, with Jerusalem in the centre and the Garden of Eden towards the edge. The map was displayed on the wall of a choir aisle in Hereford Cathedral but was apparently little regarded. In times of trouble – during the Civil War and the Commonwealth in the 17th century, for example – it was hidden beneath the floor of the chantry. During the Second World War the map was hidden again, before being returned to the cathedral in 1946.

A new library to house the map was built with public subscription and large donations from the National Heritage Fund and Paul Getty, which was opened in 1996.*

Despite the cultural revision encouraged by the Christian faith, some classical works still survived across the empire of Charlemagne, which extended across northern and western Europe in the 8th and 9th centuries. Meanwhile, in the Islamic world, a huge amount of classical learning and geography had been preserved by scholars from Baghdad to Córdoba. I call

---

* The map, the cathedral and the town are all well worth a visit. Hereford is also the birthplace of 17th-century actress and Charles II's mistress Nell Gwynn. I enjoyed a very happy couple of days there, lubricated by good cider.

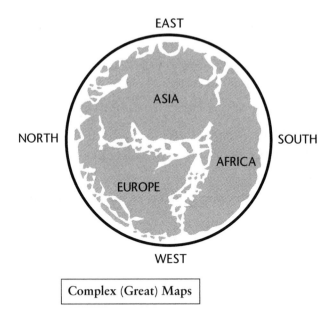

EAST

ASIA

NORTH

SOUTH

AFRICA

EUROPE

WEST

Complex (Great) Maps

Map 14. These maps show coastal details, mountains, rivers and cities, and may sometimes include figures and stories from history and the Bible.

this the Islamic tradition in geography, but it is a very loose description; there was a degree of learning and exchange between this tradition and neighbouring ones; for instance, the Christian tradition and, to a lesser degree, Indian tradition to the east. Al-Mas'udi was a fine proponent of Islamic cartography.

Al-Mas'udi (896–956 CE), referred to by some as the Arabic Herodotus, was born in 896 CE in Baghdad, descended from Abdullah Ibn Mas'ud, a companion of the Prophet Muhammad. His education took place in and around Baghdad and Basra, learning from well-known and much-admired local literary scholars. He was also influenced by the Mu'tazila school of Islamic theology, which taught that good and evil

46

were not determined by interpretations of revealed scripture, but could be established through 'unaided reason', because knowledge was derived from reason, the final arbiter in distinguishing right from wrong.

Al-Mas'udi was thus well equipped to describe and interpret the world he observed in the mid-900s CE and we owe much to the detailed observations he left us as he travelled around the Near East and Persia, North Africa, the Arabian Sea and further afield to the East African coast and to India and Sri Lanka. He took an interest in European affairs and studied the political goings on in Byzantium in particular. In his writings on the latter, in and around the year 947 CE, he uses the name 'Istanbul', rather than the usual Constantinople, five centuries before the Ottoman conquest of the city. He was aware of faraway Anglo-Saxon England and the Frankish kingdom with its capital at Paris, listing their kings up to his own time.

Travelling far to the north, he describes the Rus, an account based on personal experience and contacts made on his explorations. He describes the diverse nature of the Rus and the apparent absence of a central authority, with power in the hands of a collection of local rulers instead; they were, as he notes, capable sailors, navigating both their extensive river systems and the open sea. He was well aware that the Black Sea and the Caspian Sea were separate bodies of water, as you can see on his world map.

East of the Rus were the lands of the Kazars, a Khanate made up of a Turkic ruling elite and other diverse ethnic groups who followed various religious traditions. According to Al-Mas'udi, the Kazar rulers converted to Judaism sometime in the 740s CE and encouraged the population to follow their

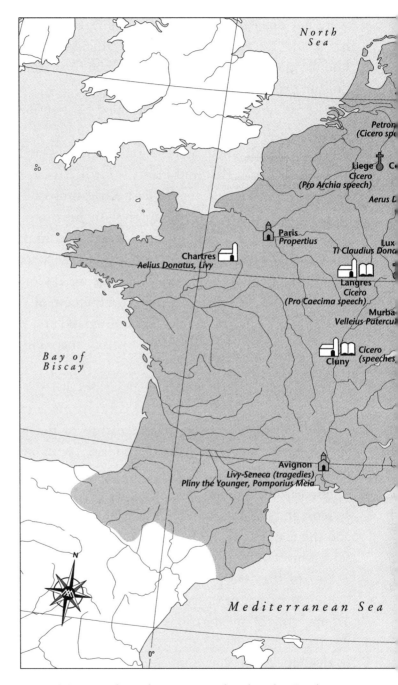

North
Sea

Petron
(Cicero spe

Liege
Cicero
(Pro Archia speech)

C

Aerus L

Paris
Propertius

Lux
Ti Claudius Dona

Chartres
Aelius Donatus, Livy

Langres
Cicero
(Pro Caecima speech)

Murba
Velleius Patercul

Bay of
Biscay

Cicero
(speeches

Cluny

Avignon
Livy-Seneca (tragedies)
Pliny the Younger, Pomporius Meia

N

Mediterranean Sea

0°

Map 15. *Some classical texts survived within the Carolingian Empire and many surviving texts were taken to the city of Florence.*

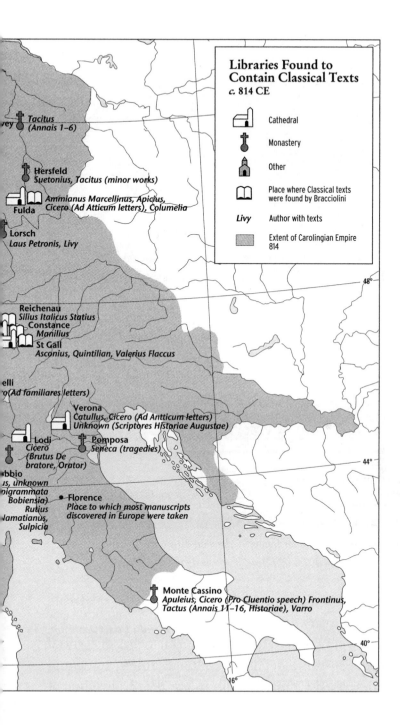

Libraries Found to Contain Classical Texts
c. 814 CE

⛪	Cathedral
✝	Monastery
🏛	Other
📖	Place where Classical texts were found by Bracciolini
Livy	Author with texts
▨	Extent of Carolingian Empire 814

Tacitus
vey (Annais 1–6)

Hersfeld
Suetonius, Tacitus (minor works)

Ammianus Marcellinus, Apicius,
Fulda Cicero (Ad Atticum letters), Columelia

Lorsch
Laus Petronis, Livy

Reichenau
Silius Italicus Statius
Constance
Manilius
St Gall
Asconius, Quintilian, Valerius Flaccus

elli
o (Ad familiares letters)

Verona
Catullus, Cicero (Ad Antticum letters)
Unknown (Scriptores Historiae Augustae)

Lodi **Pomposa**
Cicero Seneca (tragedies)
(Brutus De
bratore, Orator)

bbio
s, unknown
pigrammata
Bobiensia)
Rutius
Namatianus,
Sulpicia

● **Florence**
Place to which most manuscripts
discovered in Europe were taken

Monte Cassino
Apuleius, Cicero (Pro Cluentio speech) Frontinus,
Tactus (Annais 11–16, Historiae), Varro

48°

44°

40°

16°

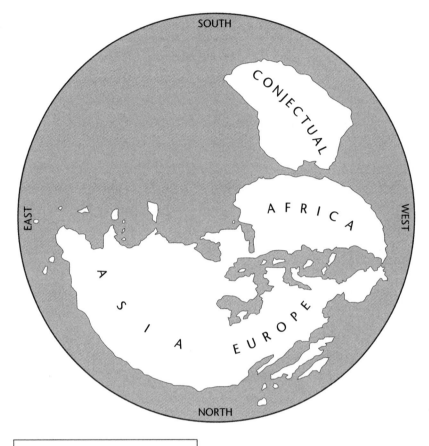

World According to Al-Ma'sudi
10th century CE

*Map 16. Al-Mas'udi was an Arab historian, geographer and explorer.
His world map, which followed Arab tradition at the time,
viewed the world with south at the top.*

example; what success they had still seems open to debate
among today's scholars.

Al-Mas'udi went on to describe the Turkic tribes of Central
Asia and wrote in some detail on India, its rulers, trade and
beliefs. He described China, though in less detail, dwelling at

one point on a revolt towards the end of the Tang period, led by a certain Huang Chao, which he describes as leading to a weakening of the dynasty.

Al-Mas'udi was a prolific writer, but the best known of his works is *Muruj adh-dhahab wa ma'adin al-jawahir* ('Meadows of Gold and Mines of Gems'), a history that begins with Adam and Eve and takes the reader through the ages to the time of the late Abbasid Caliphate. The same book mentions the story of Khashkhash Ibn Saeed Ibn Aswad, a navigator of al-Andalus (in southern Spain) who sailed across the Atlantic in 889 CE and eventually returned loaded with booty. This story is said to be well known among the people of al-Andalus – a tantalising tradition for succeeding generations of Iberians staring towards that vast western horizon.

He was an expert mineralogist and geologist, proposing a theory of evolution based on his observation of fossils; thus – from minerals to plants to animals to man – his work preceded Darwin by 900 years.

*   *   *

I try to spend as much time as I can these days in my little place in Normandy, in the middle of the Cotentin Peninsula – a great place to rest and recuperate. It is an ancient small farm, and the building is long and thin, 25 feet wide and 105 feet long. It is about half a mile from a small village, and Axel, an old and very knowledgeable local worthy, visits from time to time – just to test my wine stock, you understand, and offer advice on anything he might notice. Like most Normans, he is very generous in that way, and I soon come to realise that the shelf I have just put up in the kitchen is in the wrong place.

The Normans are an unusual group of people, an amalgam of long-settled farming folk with layers of Franks and later Vikings, the latter in particular. My neighbours, Axel included, lay specific claim to being of Viking descent – an important element of their family inheritance. In Axel's case, with a face of sun-bleached leather, white bristly hair and bright, clear blue eyes, he might just be right.

Not far from the farm lie the ancestral lands of the Hauteville dynasty. This particular Norman family arrived in southern Italy in 999 CE as mercenaries on behalf of Byzantine and Lombard overlords, and it did not take long for them to see the opportunities the area presented. On hearing the good news, more Normans, well armed and upwardly mobile, were soon on the way, eager to carve out estates for themselves. Unlike the conquest of England, which took less than a decade to consolidate, in Italy and Sicily the process took over 100 years. Eventually the Normans became local lords, and then 'the power' in the land.

King Roger II inherited his Italian and Sicilian domains from his father Roger Guiscard, Roger I, Count of Sicily, and in 1130, he established the Kingdom of Sicily. For 24 years he ruled over an unusually diverse state, made up of native Sicilians and south Italians, Arabs, Lombards, Byzantines and Normans, and fought a complicated struggle to control his lands, but in doing so he developed an amazingly inclusive and tolerant society.

Among the citizens of this realm was a certain Abu Abdallah Muhammad Ibn Muhammad Ibn Abdallah Idris al-Sharif al-Idrisi, a name that rolls off the tongue, but let's call him al-Idrisi for short. Al-Idrisi spent much of his time in the Christian kingdom of Sicily, and consequently the geographic traditions of Islam and Christianity came together in his work. He was born in the North African city of Ceuta, spent part of his early life in al-Andalus and travelled through France and into England, stopping in London and York. I was in York not so long ago, and with an afternoon to spare I walked around the area of the minster and pondered the fact that one of my cartographic heroes had, at least according to some, visited the city. With this thought on my mind I wandered into a fine little place with good beer called The Hole in the Wall, and chose the special of the day, a hearty plateful. It seemed an ancient place located on a medieval plot. What, I wondered, had al-Idrisi chosen from the menu 885 years earlier? What was the special of the day in 1130? He would have seen Normans strutting through the lanes, as they had conquered the place 60 or so years before. His later journeys took him east to Anatolia and North Africa, where he gained first-hand knowledge of those regions. Al-Idrisi chose to settle in Sicily in around 1136 and his skills eventually came to the attention of the royal household. Roger II's administration was run by a royal chancery, and among its talented staff were Greek, Arabic and Latin scribes who were capable of producing documents to suit the needs of his state and its population, in whatever appropriate language might be needed.

Around 1144, Roger commissioned al-Idrisi to produce what we might now call an atlas. King Roger II had a keen interest in the extent of his domains, which are shown at their maximum reach, in around 1150, drawn in the dark tint on our rendition of al-Idrisi's world map. In Arabic the work was called *Kitāb nuzhat al-mushtāq fī ikhtirāq al-āfāq* or 'Entertainment for He Who Longs to Travel the World'. Al-Idrisi had produced a collection of 70 detailed maps covering the known world, their orientation with south at the top. (If you look at al-Idrisi's work on the internet, his maps are frequently upside down, with north at the top, to help the modern viewer.) The atlas became known as *The Book of Roger*, indicating the king's important role in sponsoring the work. It was created following the Greek tradition, dividing the world into seven climes and moving from west to east, describing each clime in detail. Al-Idrisi referred extensively to previous Arabic geographers in his final compilation. He also relied on his own observations as well as collating the reports of contemporary travellers, many of whom, as you might expect, passed through the ports and towns of Roger's kingdom, located as it was in the central Mediterranean. The work's geographical accuracy, an improvement on Ptolemy's gigantic contribution, corrected the interpretation of the Indian Ocean as enclosed by land and the form of the Caspian Sea, and determined the direction of major European rivers. Further details were added concerning China and Tibet far in the east. In Africa, the lakes near the Mountains of the Moon are shown as the source of the Nile, whose course is tracked to the Mediterranean. The Niger River is also shown, locating the trading city of Timbuktu on the edge of the known world.

The map's style and draughtsmanship and the choice of colour are a joy to behold. They have a kind of rhythmic quality; to my eye there is a modern feel – perhaps David Hockney's pool paintings are suggested. I could stare at them for hours. Among all the triumphs of cartographic design, a calculating mind was at work on this map, as al-Idrisi estimated the circumference of the earth to be some 22,900 miles, which was just 8 per cent off from a modern calculation.

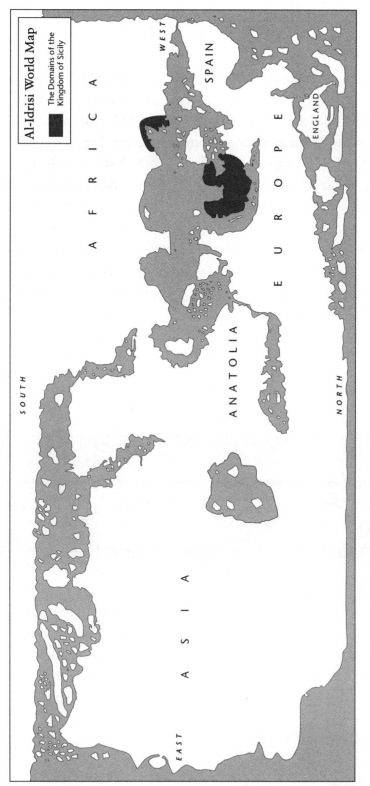

Map 17. Muhammad al-Idrisi was a geographer and cartographer. For much of his life he lived in Palermo, Sicily, at the court of King Roger II.

Al-Idrisi's work remained a standard for accuracy for almost 300 years after it was written. Geographers and travellers from Islamic traditions, such as Ibn Battuta, Ibn Khaldun and Piri Reis, were influenced by it in creating their own views of the world, while European explorers such as Vasco da Gama and Christopher Columbus also read his work in detail, thus shaping Christian tradition also.

5

VISIONS OF A NEW WORLD

Sometime just before October 1451, Cristoffa Combo was born in the Republic of Genoa, now part of modern Italy. In Italian, he was known as Cristoforo Colombo, the son of Domenico Colombo and Susanna Fontanarossa, weavers and people of the middling sort. They also had a cheese stand where young Cristoffa helped out. In 1473, Cristoffa was placed as an apprentice business agent for trading families in the port of Genoa. From there he made various voyages around Genoese holdings on the Mediterranean, which may have included a journey to Chios, a major Aegean island and trade centre, then under the control of the Republic of Genoa.

Three years later, in the spring of 1476, he served on a ship that formed part of a major Genoese convoy carrying valuable cargo to northern Europe. His particular ship was known to have docked in England, at Bristol, and also visited Galway, in Ireland. It is rumoured that in 1477, he reached Iceland. No doubt while in Iceland he would have heard rumours of Christian communities in Greenland and, indeed, of lands further to the west. In the late 1440s, the Pope had written to the Icelandic Christians concerning the welfare of their fellow believers further to the west in Greenland.

By the end of 1477, Cristoffa was known to have been on a Portuguese ship that sailed from Galway to Lisbon, Portugal. There, he met up with his brother, Bartolomeo, and together they continued trading on behalf of notable local families. Bartolomeo also worked in a cartographic workshop, which was another source of information for Cristoffa, now styled as Cristóbal Colón, who worked out of Lisbon from 1477 until mid-1485.

In 1479, Cristóbal married Filipa Moniz Perestrelo, a woman from a notable Portuguese family, and lived on Porto Santo Island in the Madeira group of islands. While Cristóbal was climbing the social ladder and somehow finding the time to learn his trade as a captain and navigator, events on the other side of the Mediterranean had changed the view of the world as seen from Lisbon. In 1453, the expanding Ottoman Empire had taken Constantinople and moved across the land route to Asia, making it much less accessible, especially for the Christian West. Meanwhile, Portugal and Spain were in the final phase of driving out the Moors from Granada, the last Islamic outpost on the Iberian Peninsula. This made the development of a new sea route around Africa to Asia much more important. Around 1470, Paolo dal Pozzo Toscanelli, a Florentine astronomer, proposed to King Alfonso V of Portugal a new route west from Portugal, which he said would be a quicker and more direct route to reach China, Japan and the Spice Islands (the Moluccas, a group of islands in what is now Indonesia), rather than the circuitous route around Africa. Alfonso rejected the idea and later King John II continued to develop the Portuguese route around Africa.

Meanwhile, Paolo dal Pozzo Toscanelli had been in correspondence with Cristóbal in 1474, proposing his westward route to the Indies. By this time, Cristóbal himself had amassed a considerable amount of his own information, including a number of maps, as well as practical experience of the Atlantic sea routes west of Portugal down to the Canaries, the Cape Verde islands and the nearby coastline of Africa. In his own studies, Cristóbal had learned from works by Ahmad ibn Muhammad ibn Kathir al-Farghani, also known as Alfraganus, who estimated that a degree of latitude around the equator represented 56.66 miles. However, Cristóbal did not realise that this was expressed in Arabic miles rather than the Roman miles that he regularly used in his navigations. Therefore, he created an error in his calculations, estimating the circumference of the earth at about 30,200 kilometres (18,765 miles), whereas the correct value should have been 40,000 kilometres (24,855 miles). He also compounded his error by following the perceived wisdom of the day that stated Eurasia – that is, from the western coast of Europe to the eastern coast of Asia – to be 180 degrees of longitude. Cristóbal went one better by following the estimate for Eurasia at 225 degrees of longitude, an estimate calculated by Greek geographer Marinus of Tyre sometime around 100 CE. This left in Cristóbal's calculations just 135 degrees of ocean between Portugal, heading due west, and the coasts of Cathay (China). Furthermore, off the coast of China lay Cipangu (Japan), which he believed lay far from its actual position, along with other smaller islands, including the mythical island of Antillia, which was believed to lie some 900 miles southwest of the Azores (which had recently been settled by Portugal in 1432).

Based on these assumptions, the great ocean to the west did not seem so empty after all. The distance from the Canary Islands – Cristóbal's jumping-off point to Japan – would be

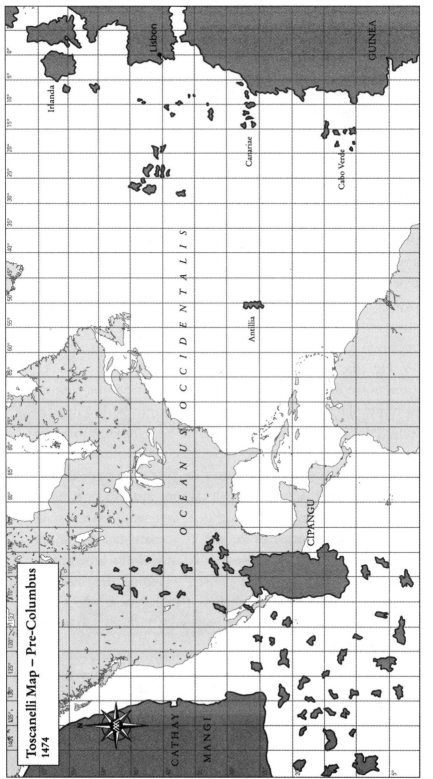

Map 18. In 1474, Paolo dal Pozzo Toscanelli created a map showing the coastline of Europe and Africa, which he mistakenly believed offered a direct route to China and the Indies.

3,700 kilometres with islands in between. It was the trade winds, becoming known as the Easterlies, that filled the sails of his small fleet, driving them across the Atlantic to his first landfall – more on that later.

In 1485, with this potential route in mind, he approached King John II of Portugal and requested the title 'Great Admiral of the Ocean', asking to be appointed governor of any lands discovered and a tenth of all revenue received from those lands in return for his modest efforts in finding a new route to Asia. Certainly, this was a step up from boy cheesemonger. King John consulted a panel of experts who looked carefully over Cristóbal's master plan and, after due consideration, recommended that the king reject the idea. Most disagreed with the calculations in degrees of distance. Cristóbal tried again in 1488 with the same plan and was rejected once more. The court was soon in a spin with the news of Portuguese explorer Bartolomeu Dias's rounding of the Cape of Good Hope in South Africa and voyage into the Indian Ocean. Cristóbal was lost in the chatter and he decided to look elsewhere.

Cristóbal first travelled to the land of his birth, the Republic of Genoa, in search of backing. He was received with a kind of polite indifference; the Genoese were focused on improving their existing trade network in the Mediterranean and Black Seas. He then moved to Genoa's great rival, Venice, where once again his representations received little sympathy. Perhaps his Genoese origins did him little service. Meanwhile his brother, Bartolomeo, had sailed north from Portugal to the court of King Henry VII of England, but this effort to gain backing was also unsuccessful.

In 1486, the kingdoms of the Iberian Peninsula had been unified by the marriage of Ferdinand II of Aragon and Isabella I of Castile who, incidentally, were reasonably close cousins who had been granted permission to marry by the Pope. It was to this royal couple that Cristóbal presented his plans on 1 May 1486. Once again, they referred to a specialist committee, which considered that the distance to Asia was vastly underestimated and stressed the impracticability of the plan to their royal highnesses, recommending against the expedition. However, in 1489, instead of outright refusing to organise an expedition, they granted Cristóbal an allowance of 12,000 maravedis per annum. To give some idea of value, this grant was equivalent to the wage of an experienced seaman. It would have paid the rent of a small cottage, with enough left over to get drunk most weekends and, if particularly frugal, pay for the occasional tattoo. Cristóbal, of course, also maintained his trading network and, at this time, is recorded as selling printed books across southern Spain, a growth industry of the period. With this grant, the monarchs at least kept their options open. They also added an order, issued to the cities and towns of their domain, to provide Cristóbal with accommodation and food at no cost.

Cristóbal continued to lobby the Spanish court, during which time he made the acquaintance of Friar Juan Pérez. The good friar had long experience as a clerk in Queen Isabella's Treasury Office. Pérez sent a letter directly to the queen informing her of Cristóbal's arrival and asking her to accept himself as Cristóbal's agent within the royal court. After two years of further negotiations, Cristóbal was eventually successful in January 1492. Ferdinand and Isabella had just completed the conquest of the last Moorish state in the Iberian Peninsula, Granada. It seems that Cristóbal may have witnessed this scene, probably after receiving comfortable accommodation and a free lunch:

After Your Highnesses ended the wars of the Moors who reigned in Europe, and finished the war of the great city of Granada where, in this present year of 1492 on 2 January, I saw the royal banners of Your Highnesses planted by a force of arms on the towers of the Alhambra.

Columbus's journal, 1492

At a meeting in the Alcázar Castle in Córdoba, Isabella had rejected his plans once again and a downcast Cristóbal was riding his mule out of town in a state of despair. However, Ferdinand intervened on hearing Isabella's description of the meeting, and immediately sent a royal official to find him and bring him back. In the second interview, Cristóbal's request was finally accepted and, in what's known as the 'Capitulations of Santa Fé', an agreement was made between Cristóbal Colón and their Catholic Majesties, and was signed on 17 April 1492, granting Cristóbal the titles of Admiral of the Ocean Sea, the Viceroy and Governor General of all new lands taken under the flag of Spain. This also included a tenth of all riches obtained from his intended voyage. Cristóbal was nothing if not persistent in his requests for such an agreement, and the elevation to Admiral would represent a considerable increase in pay from his original 12,000 maravedis. Throughout the preparations and negotiations, Ferdinand and Isabella imposed a regime of strict security, issuing a few sketchy orders to the communities and individuals involved.

Cristóbal, together with his friend and agent Friar Juan Pérez, began preparations to equip and man his small fleet in Palos de la Frontera. In order to fulfil their share of the contract, Ferdinand and Isabella issued orders instructing the town of Palos to provide Cristóbal with two ships for a period of one year, which turned out to be two caravel sailing ships named *Nina* and *Pinta*, owned by members of the Pinzón family. Cristóbal, for his part, put together his fleet of three ships and at his own cost chartered a cargo carrier named *Santa Maria*, which was owned by Juan de la Cosa, who would accompany his ship on the voyage. The total cost of Cristóbal's fleet was some 4 million maravedis, of which Cristóbal and his financial partner, the Florentine merchant Juanoto Berardi, provided 500,000 maravedis. Cristóbal persuaded members of the Pinzón family to provide captains for the fleet, and the languages spoken within the fleet were Portuguese, several dialects of southern Spain, Basque (mainly spoken by the men from the *Santa Maria*), north Italian and Latin. To this mixture Cristóbal added two translators of Arab origin, assuming that in sailing westwards he would reach the eastern lands of the Indies and Cathay. Instead, he actually bumped into an entirely new world.

Cristóbal's small fleet slipped their moorings on 3 August 1492. They headed southwest towards the Canary Islands, the westernmost Spanish possession. He dropped anchor off Santa Cruz de Tenerife, where he was delayed for some four weeks by the need for a refit and by calm winds. He eventually left the island of La Gomera on 6 September 1492. However, his ships were becalmed again within sight of the western island of El Hierro until 8 September when, once again, they could make headway. Cristóbal had calculated that the voyage should take about four weeks; however, that estimate came and went without the sighting of any land. The crews of all the ships became restless as the onboard water and food supplies ran down to the halfway point. Some of the more nervous souls argued for a return to Spain, so Cristóbal made a deal with his crews on 10 October that if no land was sighted within the next three days they would turn back for Spain.

On 12 October, just two hours past midnight, land was sighted from aboard the *Pinta* by a sailor called Rodrigo de Triana. The captain signalled the sighting by a cannon shot to Cristóbal on the *Santa Maria*, who claimed to have spotted a light on this possible landfall two hours previously. We should note here, however, that the thoughtful king had provided a small pension for the first man to sight land across the ocean; no doubt thinking of the amount of money he had borrowed from various backers, Cristóbal claimed the pension for himself and the hawk-eyed Rodrigo de Triana got nothing.

The next morning the fleet dropped anchor off the island and Cristóbal went ashore with a small landing party. He named the island San Salvador, though the natives knew it as Guanahami. The exact identity of this island is still in dispute, but the most likely candidate is one of the Plana Cays in the Bahamas. While on the island, Cristóbal met, and traded with, Native Americans belonging to the Lucayan tribe, alas now extinct. He took several members of this tribe prisoner to act as guides. After taking on water and such provisions as he could obtain, he set sail two days later. Over the next two weeks he explored a number of nearby islands, which he named Santa Maria de la Concepción, Fernandina and Isabella. These are known today as the Crooked and Acklins Islands, Long Island and Fortune Island. Before leaving the Bahamas, he visited the Ragged Islands, which he named Islas de Arena. Acting on the directions given by his native guides, he eventually arrived at Bariay Bay, Cuba, on 28 October, which he named Juana.

Thinking that he had possibly arrived on the coast of Cathay (China), Cristóbal spent many weeks in search of the Chinese

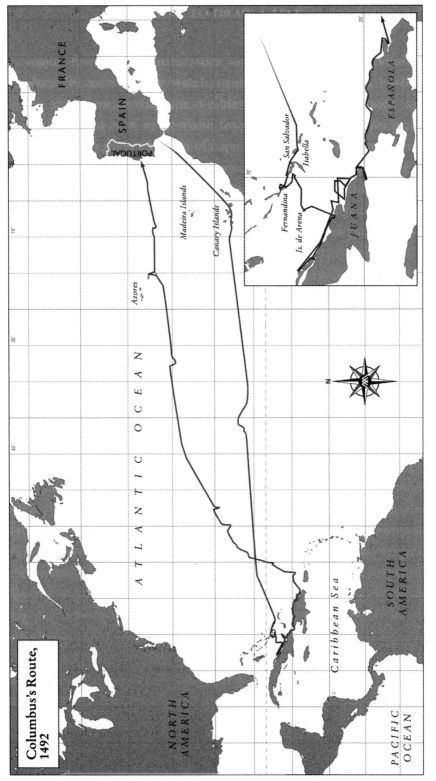

Map 19. Columbus always believed he had found a way to the East Indies and that China lay just beyond.

civilisation familiar to him from the works of Marco Polo. He coasted as far west as Cayo Cruz by 31 October. Here, north winds, together with a sense of utter frustration, instigated a change of plan. He had learned from his kidnapped native guides that gold might be found on another island further to the east. Therefore, he reversed course, sailing back along the north coast of Cuba. On 22 November, the *Pinta*, under the command of Martín Alonso Pinzón, left the fleet without permission, setting off on a search under the direction of his own native guide, looking for an island, possibly called Babeque, where he had been told that large amounts of gold could be found.

Meanwhile, Cristóbal continued his explorations with the *Santa Maria* and *Nina*, eventually arriving at the island of Hispaniola (which today comprises the two states of Haiti and the Dominican Republic), which he called Española, on 5 December. However, the flagship *Santa Maria* ran aground on a reef near Cap-Haïtien on Christmas Eve, sinking the next day. Cristóbal got his crew ashore without loss, and used the remains of his ship to build a fort on the nearby shore, which he named La Navidad (Christmas). It immediately became clear that the small caravel, *Nina*, would be unable to hold all the combined crews of the two ships. This forced a difficult decision to leave behind around 40 men to await Cristóbal's return from Spain. These men stood on the beach on 2 January 1493 and watched as the *Nina* sailed away for Spain.

Meanwhile Pinzón, in the *Pinta*, eventually discovered a number of gold nuggets in the bed of a local river (its location is disputed). Pinzón then sailed due south towards Hispaniola where, much to his surprise, he came upon the departing *Nina* on 6 January. Cristóbal's annoyance at Pinzón's behaviour eased somewhat on hearing about his discovery of gold and together they sailed for Spain, departing from Samaná Bay in the Dominican Republic on 16 January, though the ships were separated during a storm in the North Atlantic on 14 February, each commander believing the other was lost. Cristóbal sighted the Azores the next day. After a cool reception on the part of the local Portuguese governor (Portugal and Spain were not on the best of terms at this time), Cristóbal resupplied and sailed first into Lisbon on 4 March. He finally sailed back to his home port of Palos de la Frontera on 15 March 1493.

However, Pinzón, in the *Pinta*, had survived and sailed south of the Azores and eventually arrived at the port of Bayona in northern Spain. Here he made immediate repairs to his ship and the *Pinta* then sailed into Palos de la Frontera only a few hours after the arrival of the *Nina*. Pinzón fully expected to be proclaimed a hero, but this honour had already been granted to Cristóbal – literally hours earlier. Pinzón died, possibly of syphilis, a few days later, though some might argue it was due to extreme frustration and annoyance instead.

On arriving back in Spain, Cristóbal immediately wrote a letter announcing his discoveries to King Ferdinand and Queen Isabella. The letter, written in Spanish, was also copied and sent to Rome, where it was translated into Latin and printed for circulation. This letter, an excerpt of which is shown below, spread the news of the fleet's discoveries to the capitals of Europe and is perhaps the most important letter written in European history:

I have determined to write you this letter to inform you of everything that has been done and discovered in this voyage of mine.

On the thirty-third day after leaving Cadiz I came into the Indian Sea, where I discovered many islands inhabited by numerous people. I took possession of all of them for our most fortunate king by making public proclamation and unfurling his standard, no one making any resistance. The island called Juana, as well as the others in its neighbourhood, is exceedingly fertile. It has numerous harbours on all sides, very safe and wide, above comparison with any I have ever seen. Through it flow many very broad and health-giving rivers; and there are in it numerous very lofty mountains. All these islands are very beautiful, and of quite different shapes; easy to be traversed, and full of the greatest variety of trees reaching to the stars ...

In the island, which I have said before was called Hispana [Hispaniola], there are very lofty and beautiful mountains, great farms, groves and fields, most fertile both for cultivation and for pasturage, and well adapted for constructing build-ings. The convenience of the harbours in this island, and the excellence of the rivers, in volume and salubrity, surpass human belief, unless one should see them. In it the trees, pasture-lands and fruits [differ] much from those of Juana. Besides, this Hispana abounds in various kinds of species, gold and metals. The inhabitants ... are all, as I said before, unprovided with any sort of iron, and they are destitute of arms, which are entirely unknown to them, and for which they are not adapted; not on account of any bodily deformity, for they are well made, but because they are timid and full of terror ... But when they see that they are safe, and all fear is banished, they are very guileless and honest, and very liberal of all they have. No one refuses the asker anything that he possesses; on the contrary they themselves invite us to ask for it. They manifest the greatest affection towards all of us, exchanging valuable things for trifles, content with the very least thing or nothing at all ... I gave them many beautiful and pleasing things, which I had brought with me, for no return whatever, in order to win their affection, and that they might become Christians and inclined to love our king and queen and princes and all the people of Spain; and that they might be eager to search for and gather and give to us what they abound in and we greatly need.

This was a critical moment in European history and for the history of what became known as the New World. If you take a look at the map of Eurasia around 1500, you will see that standing between Europe and Asia was the growing power of the Ottoman Empire. The coasts of North Africa were occupied by warlike corsairs, the route across land through Russia was little understood and Europe was contained on its relatively small peninsula. The discovery of an unfettered sea route to the west would have a cataclysmic effect on both the peoples of Europe and the New World.

Cristóbal sailed to the 'Indies' three more times. Before setting off on his second voyage, with a fleet of 17 ships on 24 September 1493, he had received clear instructions from Ferdinand and Isabella to treat the local people with care. He arrived on the shore of Dominica on 3 November, and from there he explored the Lesser Antilles, reaching Puerto Rico on the 19th, and sailing on to Hispaniola. He arrived at the settlement of La Navidad, where he found the place empty and abandoned. The men he had left on his first voyage had, it seemed, come to blows with the locals and lost – Cristóbal concluded it was the work of the warlike Carib tribes, who had a fearsome

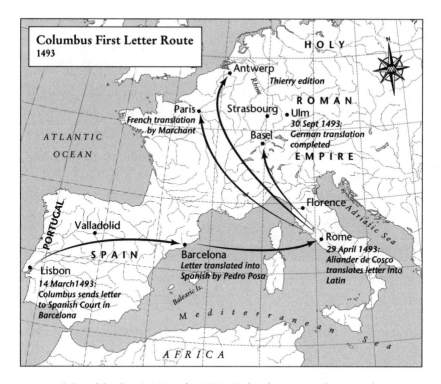

Map 20. On 14 March 1493 Columbus sent a letter to the Spanish Royal Court, then in Barcelona. Within a few months the news of his discoveries had spread across Europe.

reputation, rather than the more peaceable Taino people, with whom he was more familiar.

He moved eastwards, establishing a new settlement that he named La Isabella, which proved short-lived. He explored the interior of Hispaniola, eventually founding a fort in the centre of the island. This became a more successful location, and some gold was found there, but never enough to pay Cristóbal's backers. He forced each indigenous male over the age of 14 to deliver a quantity of gold every 12 weeks on pain of mutilation. Many failed to meet their quota and suffered hands, ears and noses being cut off; most did not survive this callous treatment. Cristóbal had established a new reputation for cruelty and exploitation and the news spread back across the Atlantic to the Spanish court. Forgetting his instructions from his royal sponsors to behave with consideration, he sent them a letter proposing to enslave the local people, notably the Carib tribe, who had demonstrated an independent spirit. This request was refused, but Cristóbal went ahead nevertheless, taking 1,600 people into captivity, mostly from the Arawak tribe. Around 560 of these were shipped to Spain, with 200 or so dying en route, while the rest were in poor condition when they finally arrived.

By exploiting and manipulating the tribes and setting them against each other and their Spanish masters, Cristóbal had pursued exactly the opposite policy to that requested by Ferdinand and Isabella. The climate and conditions in this new world were just as hard on the Spanish settlers, with losses of 50 per cent or more to fighting and disease. Leaving his brother Bartolomeo in charge of Hispaniola, Cristóbal sailed for Spain on 10 March 1496, arriving on 11 June 1496.

His third voyage, a much-reduced expedition of just six ships, sailed on 30 May 1498, after calls at Madeira and the Canary Islands, where he sent six supply ships direct to Hispaniola before he sailed to the Cape Verde Islands. After a false start, he eventually left on 22 July, steering for Dominica, and after adjusting his course north-east, he sighted an island he named Trinidad. There his ships found fresh water, before moving westwards, where they found the exit of the vast Orinoco River. Fresh water of this volume could only come from a large landmass. Cristóbal and his crew were the first Europeans to see and set foot in South America – not that it was called that then.

Cristóbal eventually returned to Hispaniola on 19 August, where he found that his brother had abandoned La Isabella and established a new settlement on the island at Santo Domingo. Conditions in the Spanish colony were not good. Meanwhile, back in Spain, numbers of returned settlers had accused the Colón brothers of cruelty and mismanagement. The king and queen decided to send in a royal administrator, Francisco de Bobadilla, who arrived on 23 August 1500. He soon took control, and the Admiral Cristóbal was required to hand over all personal and royal property on the island. His rule had come to an unsatisfactory end and Cristóbal and his brothers were sent in fetters to Spain. Back in Cádiz, Cristóbal pleaded his case, but it was six weeks before they were released and some time later before they were allowed into the royal presence. After much pleading, the brothers, to some extent, were forgiven; their private property was returned, but none of their powers.

Other explorers and merchants were allowed to sail to the west. The news that Vasco de Gama had returned to Portugal in 1499 after reaching Asia by sailing around Africa and back added insult to injury, and despite his reduced prestige, Cristóbal somehow convinced their majesties to fund a fourth voyage. He departed the king and queen's company on 14 March 1502 with the king's strict orders not to stop at Hispaniola and to concentrate on finding the mainland of India or China ringing in his ears. This would turn out to be his last and most challenging voyage.

Cristóbal set out from Cádiz on 11 May 1502 with four ships and 140 men under his command. Among the crew were his brother Bartolomeo and his younger son, 13-year-old Fernando. By now Cristóbal was 51 years old and not in the best of health. On his way, he called at Arzila, on the Moroccan coast, to rescue a band of Portuguese soldiers who had been besieged by a Moorish force. He then sailed on, arriving off the coast of Martinica (Martinique) on 15 June, his fastest transatlantic passage. He knew, by experience of the region, that a hurricane was a probability and sailed westwards, arriving at Santo Domingo on 29 June, despite his instructions not to land on Hispaniola. The new governor, Nicolás de Ovando, refused Cristóbal entry to the port and also rejected his advice on the coming storm; the first treasure fleet of 30 ships was just about to sail for Spain. Cristóbal took his four ships further down the coast, finding shelter in the river estuary of the Jaina, where he rode out the predicted storm. Meanwhile, ignoring the warning, the treasure fleet sailed, meeting the hurricane in the open sea; 29 of the 30 ships went down, with over 500 lives and tons of cargo, including a large shipment of gold, lost to the storm. One ship survived – a lucky outcome for Cristóbal, as it was carrying his own personal property and his 10 per cent share of the gold of the Indies. At least this would ensure a reasonable retirement, if he survived.

Having recovered from the storm, Cristóbal sailed to Jamaica, then on to Central America, arriving on the coast of what is now Honduras, in the Bay Islands, on 30 July. While there, he found local traders with a very large canoe, which he described as being as 'long as a galley', loaded with goods. Over the next 10 weeks he explored the coast, moving slowly south and east, arriving at Almirante Bay, in Panama, on 16 October. Over the next three months he explored the neighbourhoods of Panama in great detail. He learned from the local people of another large ocean not far from his position on the Caribbean coast. The irony of Cristóbal's investigations is that he did not press inland across the isthmus of Panama – at some stages in his explorations he must have been less than 30 miles from seeing the Pacific Ocean. In January 1503 he constructed a settlement at the mouth of the Belén River, though soon after its establishment the local Indians attacked. After restoring order, Cristóbal decided to leave the area on 16 April, heading back to Hispaniola. On the way his ships suffered storm damage and the effects of shipworm boring into their timbers, but on 10 May Cristóbal sighted what is now the Cayman Islands, which he named Las Tortugas after seeing numerous sea turtles swimming around the islands.

While approaching the coast of Cuba, his ships sustained further damage in a storm and, unable to make further passage northwards, Cristóbal sailed southwards and beached his ships in St Ann's Bay, Jamaica, on 25 June 1503. They were effectively stranded. The crews rescued whatever useful materials they could from the ships and created a shanty settlement that would be their home for a year.

Cristóbal managed to maintain a reasonable relationship with the local population, who agreed to supply him with food, although after some friction between them and his men, this food supply dwindled to almost nothing. However, Cristóbal rescued the situation by predicting a lunar eclipse that would occur on 29 February 1504. He used astronomical charts made by Abraham Zacuto, royal astronomer to King John II of Portugal, which were rescued from the wreck of his ship.

Meanwhile, a volunteer named Diego Méndez and two locals took a canoe and paddled all the way to Hispaniola where they were able to inform the governor, Nicolás de Ovando, of Cristóbal's precarious position. De Ovando detested Cristóbal and was in no hurry to effect a rescue, but help finally arrived on 29 June 1504. Cristóbal and his crew arrived back in Spain, at Sanlúcar, on 7 November 1504, concluding Cristóbal's fourth voyage.

After arriving back in Spain, Cristóbal was in a very frail condition, suffering from the effects of exposure, rheumatism and probably several other disorders. He recuperated at the monastery of Las Cuevas in Seville. Until his death, some 18 months later, Cristóbal believed he had found the Indies and the outer islands scattered along the coast of Asia. He continued to petition his royal master King Ferdinand to restore his titles and in May 1505 the king finally allowed Cristóbal an audience; this did not result in the return of his titles, but the king did allow arbitration on his financial claims. The old explorer's portion was finally confirmed at 10 per cent of the royal share. This would afford Cristóbal and his family a comfortable existence, yet the family was still not satisfied and continued to fight for the return of the titles due to them – a struggle that continued well beyond Cristóbal's death. On 20 May 1506, in the presence of both his sons, his brother Bartolomeo and his loyal friend Diego Méndez – the very same man who had paddled to Hispaniola – Cristóbal spoke his last words: 'Into thy hands, O Lord, I commit my spirit.' His body was buried in Valladolid but was subsequently moved several times in Spain and in the Indies, before finally reaching the cathedral of Seville, though many now dispute that this is his final resting place.

In the decades after his death, Cristóbal's name did not quite disappear from history, but it did sink into obscurity for a while. The Italian explorer and cartographer Amerigo Vespucci, who was born in Florence but sailed under the flag of Castile, gave his name to the Americas, and 'America' first appeared on

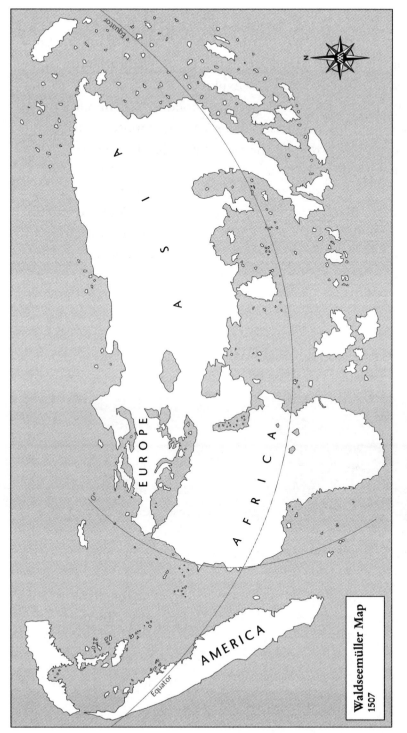

Map 21. Martin Waldseemüller published his map of the world in April 1507.
Our drawing shows the first use of 'America', after the Italian explorer Amerigo Vespucci.

Waldseemüller Map
1507

a map produced by German cartographer Martin Wald-seemüller in 1507. However, in the 1790s, Cristóbal's discoveries were reassessed and his explorations re-evaluated, to the extent that a celebrated national holiday, Columbus Day – which used his Anglicised name of Christopher Columbus – was introduced, despite the fact that Columbus never set eyes on North America.

6

PRE-COLUMBIAN VOYAGES IN THE ATLANTIC

Perhaps we should just take a moment to examine the history of transatlantic exploration. Our American cousins celebrate Columbus Day every year when we know full well that dear Cristóbal never set foot on American soil or even saw the coastline – in the interests of historical accuracy, it should perhaps be called Herjólfsson's Day (Bjarni Herjólfsson was the first European to see the Americas) or Erik Thorvaldsson's Day, after the first European to land there. But before we come to a conclusion, there are other contenders ...

St Brendan died in 577 but left a legacy of tales of seaboard exploration that circulated for 400 or so years after his death in manuscript form, crediting him with making a number of oceanic voyages, including a visit to the Canary Islands and the limits of the Western world. Irish monks certainly sailed past the Scottish Isles to the Faroe Islands and reached Iceland long before the Vikings, and in the 1970s Tim Severin successfully crossed the Atlantic in a replica Irish ship of the period, but alas there is no trace of an Irish landing in the Americas.

When the Moors invaded what is now Spain and Portugal in the 8th century CE, the story states that even bishops and

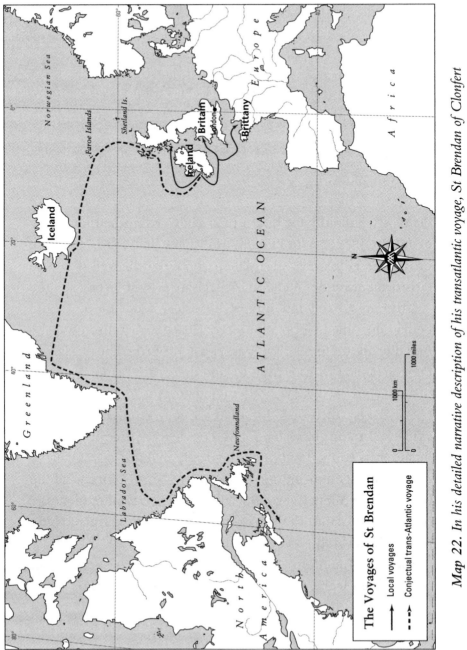

Map 22. In his detailed narrative description of his transatlantic voyage, St Brendan of Clonfert claimed to have found 'the Promised Land'.

Within the map:

The Voyages of St Brendan
— Local voyages
--- Conjectural trans-Atlantic voyage

Norwegian Sea

Faroe Islands

Shetland Is.

Britain

London

Iceland

Brittany

E u r o p e

A f r i c a

Iceland

ATLANTIC OCEAN

Greenland

Labrador Sea

Newfoundland

N o r t h A m e r i c a

0   1000 km
0   1000 miles

their flocks fled across the great western sea to establish the Seven Cities or Islands to the west. The tale flourished in Portugal and these features appeared on charts made in the late 15th century, just prior to the departure of Columbus. However, nothing was ever discovered of the Seven Cities or Islands over the following 200 years, or has been since for that matter.

My final contender is Madog ab Owain Gwynedd. According to medieval poems, he was a Welsh prince and sea rover who sailed the great sea to discover new lands. However, it seems his exploits were rewritten by Queen Elizabeth I's astrologer John Dee and the exploration writer Richard Hakluyt, in order to legitimise English claims to North America – so in fact, there were no Welsh footsteps on the New World.

This now brings us to the Vikings. Erik Thorvaldsson, better known as Erik the Red, had been banished from Norway in the mid-10th century, or more correctly Erik's father, Thorvald, had been banished, taking his family with him to settle in Iceland. This was all because of a matter of 'a few killings', according to the Icelandic sagas. Erik grew to manhood in his new land. However, history repeated itself when Erik had a disagreement with a neighbour, Eyiolf the Foul, who wound up dead, while our hero Erik ended up banished for three years, c. 982. He spent this time exploring a little-known land to the northwest, which he attractively named Greenland.

Erik the Red promoted Greenland and attracted numbers of settlers from Iceland and beyond. A few years after the Greenland settlements were established, Bjarni Herjólfsson, as part of a small fleet of ships carrying settlers to Greenland, was blown off course and after three days of sailing alone he sighted land to the west. However, Bjarni would not be distracted. He was intent on one thing and that was making his way to the farm already settled by his father in Greenland. On his arrival he described the lands he had seen to Leif Erikson, an Icelander and an experienced captain. Leif made an offer to Bjarni to purchase his ship, and Bjarni agreed. Leif then set about gathering a crew, which eventually numbered about 35 men.

Leif had listened carefully to Bjarni's description, taking great notice of the winds and currents. Leif sailed from Greenland, heading westward across what we now call the Labrador Sea, following the landmarks described by Bjarni. The ship he used was called *Knarr*; this was not the dragon-headed Viking ship of legend, but an altogether stouter, tubbier ship intended to carry cargo and livestock. These vessels were powered by sail and oar and displaced a shallow draft, which meant they were ideal for investigating inshore river estuaries and, if necessary, following rivers inland. The first land Leif saw he called Helluland (suspected to be the southern part of Baffin Island in Canada), which means 'land of flat stones'; as you can imagine there was not a whole lot to interest Leif in this region. He sailed on further south encountering an area he called Markland, the 'land of forests'. From an almost treeless Greenland, there was a lot to interest Lief here – good timber was a very useful resource.

At the end of 1001, Leif decided to build a winter camp, which was probably located near Cape Bauld on the northern tip of Newfoundland. Leif and his crew would be the first Europeans to explore this region until the arrival of John Cabot 496 years later. Leif spent another year and another winter at the settlement, which he called Leifsbuthir. He seems to have existed peacefully with the local people and must have

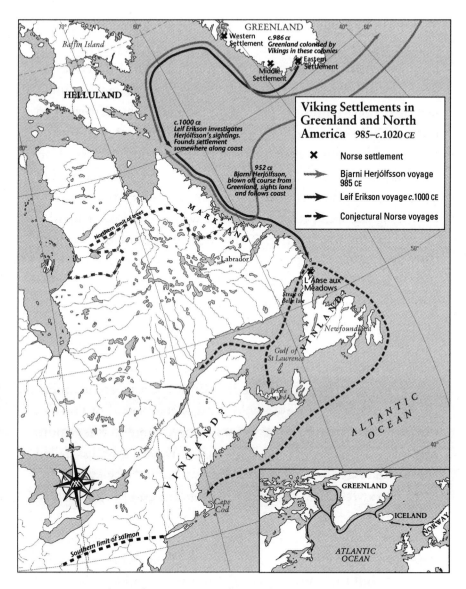

Viking Settlements in Greenland and North America 985–c.1020 CE

- **✗** Norse settlement
- Bjarni Herjólfsson voyage 985 CE
- Leif Erikson voyage c.1000 CE
- Conjectural Norse voyages

GREENLAND

✗ Western Settlement c.986 CE *Greenland colonised by Vikings in these colonies*

Baffin Island

✗ Eastern Settlement

✗ Middle Settlement

HELLULAND

c.1000 CE Leif Erikson investigates Herjólfsson's sightings. Founds settlement somewhere along coast

952 CE Bjarni Herjólfsson, blown off course from Greenland, sights land and follows coast

Northern limit of trees

MARKLAND

Labrador

L'Anse aux Meadows

Strait of Belle Isle

Newfoundland

Gulf of St Lawrence

VINLAND

Prince Edward Island

ATLANTIC OCEAN

St Lawrence River

VINLAND

Cape Cod

Southern limit of salmon

GREENLAND

ICELAND

NORWAY

ATLANTIC OCEAN

Map 23. Vikings first set eyes on the North American continent in 1004. Voyages from their settlements in southern Greenland may have continued up to the early 1400s.

explored much further south. At any rate, during the second year of his stay an older servant called Tyrker managed to get himself drunk. Tyrker clearly had time on his hands to experiment in the creation of alcoholic beverages. Berries growing in this region could obviously be fermented and Leif described them as 'Wine-berries'. Whether this could be a description of Vinland or whether that was located further south is unclear. In 1003, Leif gathered his crew and sailed back to Brattahlid in Greenland in order to fulfil duties on the family farm.

During the long, dark Greenland winter, Leif described his expedition to his brother, Thorvald. In the spring of 1004, Thorvald, with a crew of 30 men, retraced his brother's voyage and stayed over winter at the Leifsbuthir settlement. The following spring, there was a skirmish recorded with the local people, some of whom were captured, although one escaped and returned with a larger force. They fought another battle with Thorvald and his crew and the unfortunate Thorvald was struck by an arrow and killed. A brief period of hostilities continued, but the Norse explorers saw the period of troubles through and stayed again over another winter. After more explorations, they left for Greenland the following spring (1006), leaving Thorvald's body buried in the New World.

The next recorded visitor to North America was Thorfinn Karlsefni, also known as Thorfinn the Valiant. He set sail with three ships and somewhere between 160 and 250 settlers, depending on which account you read, and also livestock. These would be the first recorded European domestic animals imported into North America. Thorfinn and his settlers landed at a place called Straumfjord; he later moved and created another settlement, which he called Straumsöy, probably · changing his location for better defence. There was a period of peaceful coexistence between the Native Americans and the Norse settlers, with the latter bartering for furs and grey squirrel skins in exchange for milk and a red home-woven cloth, which proved to be highly valued by the local people and was used in their headdresses.

During Karlsefni's stay in North America there must have been a period of exploitation; the most valuable of commodities was timber and many shiploads must have been sent to the Greenland colony, where timber was in great demand. Aside from this traffic, other explorations would likely have taken place. There were certainly enough settlers, with farmers and tradesmen among them, to be able to construct and provide a reasonable lifestyle in the New World. However, we must remember this was at the furthest reaches of Norse exploration. Back in the Scandinavian homelands at around the time of Karlsefni's New World adventure, the Scandinavians were far more focused on opportunities in Europe, from the British Isles in the west to the Russian river system in the east. The Norse settlements in North America just didn't quite generate enough attention among senior Scandinavian states to supply a steady flow of settlers, even though in many ways the Norse people living in Newfoundland fitted in more naturally with the climate and resources than the early English settlers 500 or 600 years later.

Karlsefni's stay in North America came to an end after a period of disagreement with the Native Americans. The sagas describe one particular battle in which the settlement is attacked and, outnumbered, the Norsemen begin to retreat. At this point Leif Erikson's half-sister, Freydís Eríksdottir, who was about eight months pregnant at the time and unable to keep up with the retreating Norsemen, called upon them to stop running away and fight. Left alone on the battlefield, she stooped and picked up an abandoned Viking sword and turned to scream at her tormentors. An outraged pregnant Scandinavian was too much for the locals, who turned and fled, but despite this victory, Karlsefni decided to leave the settlement and return to Greenland. The Greenland colony would exist for about another 400 years, and during this time its inhabitants would have traded for timber and furs with fairly regular voyages to the New World, until they were last heard of in the 1420s.

7

THE FIRST CIRCUMNAVIGATION OF THE WORLD

Fernão de Magalhães, better known as Ferdinand Magellan, was born in Portugal in 1480. In March 1505, Magellan, then 25 years old, enlisted in a Portuguese fleet of 22 ships sailing on behalf of the Portuguese king, taking the first Viceroy of Portuguese India to take up his new post. Magellan took part in the Battle of Cannanore in 1506 where he managed to get himself wounded, and in 1509, he was again in the thick of things at the Battle of Diu. As well as becoming a skilled sailor, he knew how to survive on the battlefield.

Magellan later sailed with Diogo Lopes de Sequeira, who was leading the first Portuguese embassy to Malacca, an important trading state located on the Malay Peninsula. They arrived in September 1509, but this embassy ultimately fell victim to local politics, and there were plans to attack its leading members. By now a seasoned veteran, Magellan was able to warn Sequeira and his friend Francisco Serrão who, with his help, both escaped, and later Francisco Serrão became one of the first Europeans to pass beyond the Straits of Malacca, sailing in a small fleet of four ships on a mission to find the exact location of the Spice Islands, in what is now Indonesia.

Serrão was able to send letters back to Magellan via the extended Portuguese network informing him of his new discoveries.

Meanwhile, Magellan fell out of favour with the Portuguese authorities after taking leave without permission and he returned to Europe. He served in Morocco for the Portuguese, where he received yet another wound, which resulted in a permanent limp that lasted the rest of his life. While in Morocco, he was maliciously accused of trading illegally with the Moors, and although these accusations were later proved false, his ongoing relationship with the Portuguese authorities deteriorated. He petitioned King Manuel I to form an expedition to sail westwards from Portugal across the Atlantic, around the Americas and across the Pacific to the Spice Islands, but the request was denied.

Magellan left Portugal for Spain where he immersed himself in researching the latest available charts and maps relating to his proposed voyage. He befriended fellow Portuguese cosmographer and mapmaker Rui Faleiro, who became his advisor while he prepared. The Spanish authorities, for their part, were interested in proving that the Spice Islands might fall into their area of exploitation as defined by the Treaty of Tordesillas of 1494. Eager to prove this a possible new route to the Spice Islands, which would be free of Portuguese interference, the Spanish Crown had already sent expeditions under Vasco Núñez de Balboa, who reached the Pacific in 1513 after marching across the isthmus of Panama. This was followed by the expedition of Juan Diaz de Solis, which explored as far south as the River Plate by 1516.

Magellan and his partner, Rui Faleiro, met a senior factor of the Casa de Contratación ('House of Trade'), Juan de

Aranda, who accepted their proposal and later presented it to the Spanish king, Charles I (who would become Emperor Charles V), stating that if this expedition were successful it would be the realisation of Columbus's original plan to reach the Spice Islands by sailing westwards. Following recent discoveries, King Charles I named Magellan and Faleiro the senior captains of a new fleet that would number five ships. The deal that the two explorers received was extremely generous: it included a monopoly of the pioneering route for a period of 10 years; they would be appointed governors of the lands and islands discovered, with 5 per cent of the resulting net gains from the new territories and a fifth of all gains as a result of the journey; they would be granted the right to levy 1,000 ducats on further vessels following their pioneering route and would receive an island each – not too bad; an explorer could get on with that. The Spanish Crown largely supplied the initial funding. King Charles also raised them to the rank of Commander of the Order of Santiago, which involved a smart badge and uniform for those important social occasions.

The fleet, including the flagship *Trinidad* accompanied by the *San Antonio*, *Concepción*, *Santiago* and the *Victoria*, assembled in Brazil. The multinational crews for the ships were also assembled – mostly Spanish and Portuguese, but also including Italians, Greeks, some French, Flemish, Dutch and a couple of Englishmen. Included in this number was the indentured servant of Magellan, Enrique of Malacca, the only known Asian in the crew. The other notable crew member was Antonio Pigafetta, originally from Venice, who was a scholar and explorer; he would become a close assistant to Magellan and would also keep an accurate journal of the voyage. Juan de

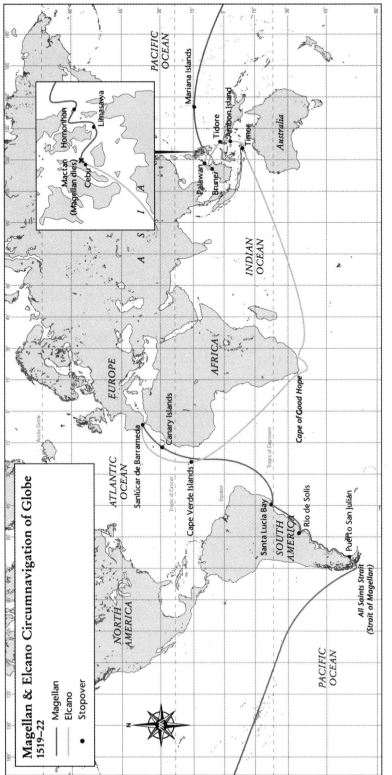

Map 24. Ferdinand Magellan departed on his global voyage on 20 September 1519. He was killed in the Philippines, and Juan Elcano took over, returning on 6 September 1522.

Cartagena was named Inspector General of the expedition and was responsible for keeping a detailed log of its financial costs and trading operations. Finally, Franco Albo also kept a formal logbook of the voyage.

On 10 August 1519 the five ships left Seville, descended the Guadalquivir River and entered the Atlantic at Sanlúcar de Barrameda, where they awaited favourable winds, finally embarking on their fateful journey on 15 September 1519. The ever-watchful King Manuel I of Portugal had received reports from his spies that his fellow countryman was now in the service of Spain. On hearing of Magellan's departure, he ordered a Portuguese naval detachment to pursue him. However, the wily navigator managed to evade the pursuing Portuguese and sailed on to the Canary Islands unimpeded. On leaving Cape Verde, he and his crew crossed the equator on 27 November and sighted the South American coast on 6 December. A few days later, on 13 December, they dropped anchor in the area of present-day Rio de Janeiro. There was no permanent Portuguese presence here, despite the fact that Pedro Cabral had claimed this region for Portugal. Magellan's fleet resupplied without interference, and eventually set sail again, reaching the River Plate (Rio de la Plata) in early February 1520. Continuing to move south, Magellan decided to over-winter and founded a temporary settlement, which he called Puerto San Julián, on 31 March 1520. While there, the crew met native people that Antonio Pigafetta described in his journal as 'giants'; he called them Patagonians, though he doesn't say how he arrived at this name.

While the crew were over-wintering, a mutiny broke out on 1 April, which involved three of Magellan's five ships: the *Victoria* under Captain Luis de Mendoza; the *Concepción*, commanded by Gaspar de Quesada and the *San Antonio*, commanded by the head of the mutineers, Juan de Cartagena. After a short but decisive confrontation, as we might expect from the battle-hardened Magellan, Mendoza was killed and the other ships were quickly retaken and brought back under control. Magellan immediately tried the mutineers and executed several of them, including Gaspar de Quesada, and he also marooned Juan de Cartagena and his priest, Sanchez de la Reina, on the coast before continuing his journey. Another mutineer, Juan Sebastián Elcano, was spared but spent five months in chains as punishment; a tough old salt, he will feature later in the story. In Pigafetta's report he mentioned that the guilty parties were 'drawn and quartered' with their remains impaled on stakes along the coastline, to be discovered by Francis Drake many years later. Obviously, Magellan was a commander you did not mess with, although he did forgive many of the ordinary seamen.

The *Santiago* was sent ahead down the coast on a scouting expedition while the remaining four ships were refitted and prepared for the forthcoming journey. Unfortunately, the *Santiago* was caught in a sudden storm, blown onto the nearby coast and wrecked. By some miracle, all the crew managed to get ashore to safety, and two of the survivors volunteered to walk overland back to Magellan's anchorage and organise a rescue. After this loss, Magellan decided to postpone his departure from Puerto San Julián for a few more weeks in the hope that the weather would improve, which in due course it did. On 21 October 1520, the four ships reached Cape Virgenes; at this spot Magellan and his commanders concluded that they had found the passage that would lead them through to the Pacific. The decision was based on the water quality and the fact that the channel led deep inland. The four ships cautiously undertook that 600-kilometre passage, which Magellan named Estrecho de Todos los Santos ('All Saints Strait'). These days the strait is called the Strait of Magellan. He had assigned the *Concepción* and *San Antonio* to take the lead; however, the *San Antonio*, under the command of Gómez, deserted, turned about and headed back to Spain on 20 November. The remaining three ships emerged from the passage, entering the Pacific Ocean on 28 November. Magellan was impressed by the peaceful and calm nature of this '*Mar Pacifico*', the Pacific Ocean. He and his crewmen were the first Europeans to reach Tierra del Fuego, the 'land of fire', on the Pacific side of the strait.

Magellan headed north, initially following the western coastline of South America, then turned northwest, heading directly across the Pacific Ocean, and reached the equator on 13 February 1521. If there was ever going to be an outbreak of nervousness among the crews of his ships, it would be out there in the vast emptiness of the Pacific Ocean. There were a few sea birds to be seen, and perhaps the odd whale going about its lonely business, but they were far from any land mass other than a few small islands. Indeed, on 6 March, they sighted what we now call Guam, one of the larger islands (at 210 square miles) in the western Pacific. The inhabitants of the island seemed more curious than wary and apparently swarmed onto the three ships, stealing everything they could lay their hands on. They also managed to make off with the ship's rowing boat, which was being towed by Magellan's flagship. By 16 March, Magellan and his squadron reached the island of Homonhon in the Philippines. His three ships still mustered a total crew of 150 men, who were the first Europeans to reach the Philippine archipelago.

Magellan made his way through the islands, arriving at Cebu on 7 April. The local ruler, Rajah Humabon of Cebu, exhibited friendly intentions towards Magellan and his crews. Both the Rajah and his queen were baptised as Christians and were given images of the Holy Child along with a cross, later known as Magellan's cross. These were regarded as symbols of the ongoing Christianisation of the Philippines. Rajah Humabon, together with his allies, wanted Magellan to kill off their enemy, a certain Datu Lapu-lapu, the ruler of Mactan. Magellan would have preferred to convert Lapu-lapu to Christianity, but Lapu-lapu rejected a change of religion for himself and his people. In the early morning of 27 April 1521, Magellan sailed to Mactan with a combined Spanish and local attack force. In the ensuing battle against Lapu-lapu's forces, Magellan, leading his men, was hit by a bamboo spear, surrounded and eventually killed with what has been described as a large cutlass. Pigafetta, in his notes, stated that 'nothing of Magellan's body survived'.

That afternoon, the grieving Rajah Humabon, hoping to recover Magellan's remains, offered Lapu-lapu a handsome ransom of copper and iron for them. Lapu-lapu refused, intending to keep the body as a war trophy. The casualties suffered in this short campaign left too few men to crew all three ships – they could only muster 115. Therefore, on 2 May, the *Concepción* was abandoned and burned. The *Trinidad* and *Victoria* left on 21 June, heading westwards, guided by local Moro pilots. After much indecision regarding Magellan's succession to command, the crews eventually elected Duarte Barbosa and João Serrão. However, within four days of their election, they were both also killed in a skirmish. The survival of the expedition now hung in the balance. One man, Lopez de Carvalho, stepped forward a little hesitantly and now took command. He proved somewhat indecisive, however, and after a long and tedious journey, the crews became more impressed by the abilities of Juan Sebastián Elcano, a Spanish explorer of Basque origin, and he emerged as their leader. The expedition arrived in Brunei 35 days after setting sail and reached the Spice Island of Maluku on 6 November. Here, they traded with the islands of Tidore and Ternate for the valuable cargo of cloves and other spices.

Once Juan Sebastián Elcano had assumed command, on 18 December, the ships made ready to leave the Spice Islands, but the *Trinidad* had sprung a leak. Carvalho stayed with his ship, along with 52 of his crew, intending to make repairs and continue later.* Meanwhile, Elcano, commanding the *Victoria*, sailed westwards with 54 European and four Asian crewmen on 21 December. By 6 May 1522, the *Victoria* rounded the Cape of Good Hope, where the crew took on new provisions with the permission of the Portuguese authorities who controlled the area, before sailing on to Spain via the Canary Islands, arriving on 20 September 1522. The *Victoria* had just 18 survivors on board; some had died en route; others chose to leave the ship in South Africa and the Canary Islands, probably due to illness. Thus, it was Elcano who completed the first, unintentional, circumnavigation of the globe. The original intention was that Magellan would find the western route to the Spice Islands and return by the same route. The king granted Elcano a coat of arms featuring a globe and the motto

---

* After making repairs, Carvalho attempted to return to Spain via the Pacific route, which failed – the *Trinidad* was captured by the Portuguese and later wrecked. So ended the attempt to complete the original plan.

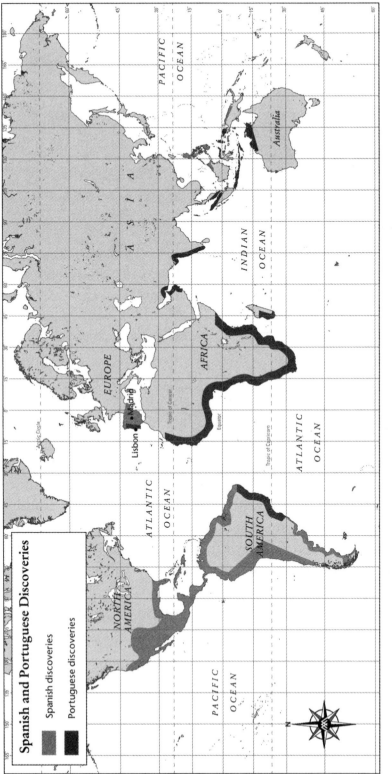

Map 25. During the course of the 16th century the Spanish discovered most of the Americas and the Portuguese most of Africa and Asia.

Primus circumdedisti me ('You went around me first'), as well as a useful pension.

*   *   *

As the experience of the voyage became understood in Spain, it was clear that another expedition was needed to provide secure access to the Spice Islands. The new expedition was duly organised, led by Garcia Jofre de Loaísa, with the mission to colonise the Spice Islands. It set sail in July 1525, becoming the second expedition to cross the Pacific Ocean, and arriving in the Spice Islands on New Year's Day 1527. This venture was based upon the belief that the demarcation line laid down in the Treaty of Tordesillas left at least part of the Spice Islands (depending on whose argument you follow) in the Spanish Zone. The treaty had effectively been an attempt to separate the world into two halves, one intended for Portuguese exploitation and the other for Spanish exploitation. In effect, the Portuguese had gained a slightly larger portion, though. Instead of the 180-degree line intended by the treaty, Portugal was actively exploiting 191 degrees and Spain was scraping by with just 169 degrees. The expedition eventually anchored off the island of Tidore where the Spanish constructed a fort and established themselves, which led to inevitable conflicts with the Portuguese who were already established on the nearby island of Ternate. There would be a decade of skirmishing between the two parties.

Meanwhile, back in Europe, the Spanish King Charles I, who was now also Emperor Charles V of the Holy Roman Empire, and the Portuguese King John III decided that some kind of treaty between the two countries clarifying the status of the Spice Islands needed to be brokered. This rapprochement was mainly initiated by Charles V, who had weighty matters on his mind – he was at war with France and the flow of gold and silver from his overseas territories needed to be protected and enhanced, and skirmishing with the Portuguese was a waste of resources. At any rate, the treasures pouring into Spanish coffers were never enough to fulfil the ambitions of Spain in Europe. The two parties organised two delegations, which would once and for all recognise the boundaries by which the two countries would exploit the world. This agreement was eventually concluded and signed on 22 April 1529 and became known as the Treaty of Zaragoza, though it is occasionally referred to as the Capitulation of Zaragoza. The map showed clearly that the Spice Islands were well within the Portuguese Zone. The Spanish moved out of the Spice Islands on payment of a large sum made by the Kingdom of Portugal, which was a useful source of finance for Charles V's wars. Meanwhile, the Spanish had been colonising the Philippine Islands and Charles rather hoped that since the Philippines contained no spices, Portugal would turn a blind eye to the fact that they lay in their zone – which indeed they did.

*　　*　　*

Meanwhile, in the thriving Spanish colony of Cuba in 1518, the governor of New Spain, Diego Velázquez de Cuéllar, had placed Hernán Cortés de Monroy y Pizarro Altamirano – an ambitious and determined individual – in command of an expedition to explore and, if possible, take possession of central Mexico.

However, due to an old dispute between the two men, Velázquez revoked his charter to Cortés, ordering him to stop the expedition. Cortés ignored the order and departed in February 1519, collecting soldiers and extra horses en route, eventually commanding a force of 11 ships, about 500 men and 14 horses.*

Cortés landed on the Yucatan Peninsula and in March 1519, he formally took possession of the land in the name of the Spanish Crown. He did this as a legal sleight of hand, having fallen out with Governor Velázquez – it would give him a direct line to Charles V back in Spain, or so he hoped. While in the area, he met a stranded priest, Gerónimo de Aguilar, who had been captured by the local Mayan people, though he had successfully escaped. While in captivity, Gerónimo had learned the local Mayan language and joined Cortés's expedition as a translator. The expedition then proceeded to the area of Tabasco, where Cortés fought a battle against the local people, winning a complete victory. The locals surrendered and, as a token of their submission, presented their conquerors with 20 young women. Cortés immediately asked his priest to convert them all to Christianity. Among this group of converts was an intelligent woman known as La Malinche, who became Cortés's mistress. She could speak the local language but also, more importantly, the language of the Aztecs and had a working knowledge of their social structure. This would all benefit Cortés after he landed his force on the coast near Río Jamapa. He then moved north through Zempoala and took Veracruz in July 1519. While there, he formally dismissed the authority of

---

* As an ordinary soldier you got a double share of the spoils if you brought a horse.

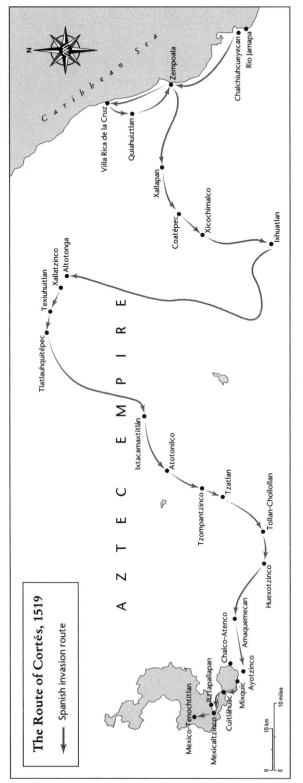

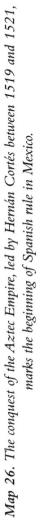

Map 26. The conquest of the Aztec Empire, led by Hernán Cortés between 1519 and 1521, marks the beginning of Spanish rule in Mexico.

Governor Velázquez and placed himself in full command in the name of King Charles V of Spain. To prevent any idea of revolt or retreat by his soldiers, he also scuttled all his ships.

While in the area of Veracruz, Cortés met the leaders of local tribes, all of whom were under Aztec rule. The locals were friendly towards the strange, bearded foreigners and declared that they were not so friendly with the Aztecs. Cortés successfully made alliances with the indigenous people, by force when necessary, and arranged a meeting with Montezuma II, ruler of the Aztec Empire, to attempt to arrive at some kind of trade agreement, particularly relating to the Spanish acquisition of gold and silver. By October 1519, Cortés and his army, now numbering over 1,000 men, marched on Cholula, the second city of the Aztecs. Upon arrival in Cholula, in a premedited effort to instil fear in the Aztec elite who were gathering to meet him at Tenochtitlan, the capital city, Cortés slaughtered most of the local leaders and burnt most of Cholula.

Once in the vicinity of Tenochtitlan, the Spanish and their allied army had grown larger. On 8 November 1519, they were received with full dignities by Montezuma, who had realised the threat they represented, allowing the Spaniards to enter the island city of Tenochtitlan. Cortés was very keen to do this as he could look around and see any potential weaknesses in their defence. Montezuma presented the Spanish with lavish gifts, including large amounts of gold – which was a big mistake, since gold meant immediate profit in the eyes of the Spanish. Rather than impressing the Spaniards with his generosity, Montezuma had encouraged their greed. It had come to the attention of Cortés that the Aztecs considered Montezuma an almost god-like figure (a view rejected by some modern historians). When Cortés seized Montezuma II as a hostage within his own palace, he set about ruling Tenochtitlan, using Montezuma as his puppet.

Meanwhile, Governor Velázquez, back in Cuba, had organised a new expedition, numbering 1,100 men and commanded by Pánfilo de Narváez, in order to challenge Cortés. They arrived in Mexico in April 1520, and on hearing this news, Cortés marched out with his own army to confront Pánfilo de Narváez, leaving 200 men behind in the city. Despite being outnumbered, Cortés overwhelmed Narváez's force, the bulk of which he persuaded to join him. Meanwhile, back in Tenochtitlan, a trusted lieutenant of Cortés, Pedro de Alvarado, had committed a massacre of local notables on 22 May 1520, on the Feast of Toxcatl, believing that he had evidence of an Aztec plot to attack and murder the Spanish. This event triggered a general rebellion among the population. Cortés hurried back to Tenochtitlan, and in the chaos of 1 July 1520, Montezuma II was killed. Given the enraged state of the population, Cortés decided to withdraw, heading for the city of Tlaxcala. During the withdrawal the fighting continued, and although the bulk of the Spaniards escaped, most of the rearguard and, even more importantly, much of the looted treasure was lost. The Spanish managed to withdraw, fighting another battle at Otumba on 7 July 1520, and finally arrived at Tlaxcala having lost 880 men. Here they regrouped, receiving reinforcements from Cuba and their native allies. Cortés then set about besieging Tenochtitlan, cutting off supplies to the city. The siege eventually ended in the complete destruction of the city on 13 August 1521. The last emperor of the Aztec Empire, Cuauhtémoc, was also captured. Cortés took over the ruins and began rebuilding, renaming the place Mexico City and claiming the area for the Spanish Crown. He would personally govern Mexico until 1524.

A year and a month after the capture of the Aztec Empire, Juan Sebastián Elcano returned to Spain, having completed the first circumnavigation of the world. It was becoming clear to the Spanish that finding a sea route to the Pacific through the New World was becoming less likely and that the only solution would be to sail around the southern tip of South America, a long and tedious journey. Now with central Mexico in Spanish hands, why not station a fleet off the west coast of Central America? With this base now established, the trade route to Manila in the Spanish-held Philippines from Acapulco was inaugurated in 1565 and lasted until 1815.

8

THE ENGLISH WORLD VIEW

JOHN CABOT

John Cabot was born Giovanni Caboto in about 1450 in Geata in the province of Latina on the western coast of Italy. He was a trader-merchant, navigator and explorer who became a citizen of Venice in 1476, allowing him to engage in the maritime trade that was concentrated in the eastern Mediterranean. However, his trading account seemed to get into trouble in the late 1480s and he left Venice for Valencia in 1488, pursued by his creditors. He left for Seville in 1494, where he hoped to gain a construction contract, but it fell through. The troubled Caboto then moved again, arriving in England by mid-1495, when he proposed an exploration in search of new lands. By 5 March 1496, Caboto (now Cabot) held 'letters patent' from the hand of King Henry VII giving permission to explore with:

> free authority, faculty and power to sail to all parts, regions and coasts of the eastern, western and northern sea, under our banners, flags and ensigns, with five ships or vessels of whatsoever burden and quality they may be, and with so many

and with such mariners and men as they may wish to take with them in the said ships, at their own proper costs and charges, to find, discover and investigate whatsoever islands, regions or provinces of heathens and infidels, in whatsoever part of the world placed, which before this time were unknown to all Christians.

John Cabot hurried to Bristol to prepare for the expedition. The port already had contacts with Iceland and was full of stories of islands out in the Atlantic and gossip that Bristol men had already found islands far to the west.

He departed some time in the summer of 1496, but his preparations were inadequate and it is said that he met with bad weather and contrary winds, which eventually forced him to abandon the voyage and return to Bristol to try again the following year.

He sailed again from Bristol in May 1497 in a single vessel named the *Matthew*. The records we have of this voyage are largely based on two letters: one written by Lorenzo Pasqualigo and another by John Day, which was sent to Christopher Columbus in 1498. Within four weeks of sailing they sighted land, but since no records survive this could have been anywhere between southern Nova Scotia and northern Newfoundland. Cabot did effect a landing, but only as far as a crossbow could fire. He discovered the remains of a campsite and a collection of animal snares set to trap local wildlife. With this information, he quickly retreated to the safety of the *Matthew*. He was concerned that the natives might lurk among the dense woodland not far off. He was also convinced, as Columbus was, that he had encountered the lands of the Emperor of China and that further inland would be great cities

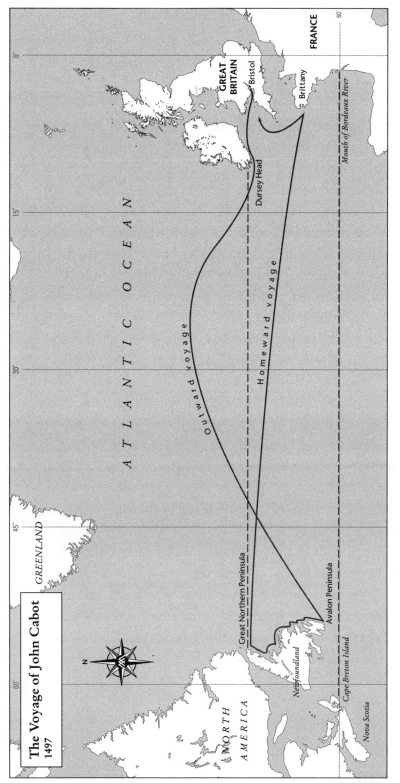

Map 27. John Cabot, also known as Giovanni Caboto, navigator and explorer, sailed on his first voyage across the Atlantic in 1497.

and wonderful trading opportunities. Also like Columbus, he was completely wrong.

However, Cabot had re-established European knowledge of the area since the decline of the Norse settlements in Greenland several hundred years previously. The information he brought back to Bristol concerning the rich fishing grounds off Newfoundland would instigate a flow of eager fishermen from England, France, Spain and the Basque country, who would come to understand the coastline of Newfoundland and Nova Scotia in detail. On Cabot's arrival back in Bristol, King Henry, who must have been slightly less impressed by Cabot's discoveries, awarded him the princely sum of £10 for what he described as a 'new isle'.

A new and larger expedition was organised for the following year. A fleet of five ships were equipped and dispatched in 1498 and it is recorded that one ship turned back, while the other four were never heard of again. Some years later, a contemporary account stated that Cabot was believed 'to have found the new lands nowhere but on the very bottom of the ocean'.

WILLIAM WESTON

Meanwhile, among the merchants of the port of Bristol, there was talk of opportunities across the Atlantic. Recent research has brought to light the name of one of these merchants – William Weston. He is known to have traded with the Portuguese island of Madeira out in the Atlantic and he was also part of a Bristol expedition aboard the ship *Trinity*, which went in search of the 'Isle of Brazil'.

Weston may have been working under the 'letters patent' provided by the king to Cabot. It was a legal possibility that these 'letters patent' could be utilised to employ a third party and that some time between 1499 and 1500 Weston did indeed undertake a voyage that again visited Newfoundland and may have explored further north along the shores of Labrador. There is evidence of a payment, approved by the king, made by two customs officers in 1500, which reads:

> Namely £30 above the fee [of the customers] and let them have for the said £30 a tally for W[illiam] Weston, merchant of Bristol, for his expenses about the finding of the new land.

SIR FRANCIS DRAKE

Sir Francis Drake was born in Devon, southwest England, c. 1540. He was firstly a sea captain, then a naval officer, explorer and slave trader. According to the English, he was also a privateer, and according to the Spanish, a pirate.

We pick up his career in 1577. Elizabeth I of England dispatched Drake on an expedition against Spanish trade along the Pacific Coast of South America. Sir Richard Grenville had already drawn up the plans for the voyage, but with political pressure from Spain they were later rescinded. With or without 'letters patent', Drake set off from Plymouth on 15 November 1577. However, bad weather forced his crew to return to Falmouth, where they refitted and set off again on 13 December with five ships and 164 men. He later added a captured Portuguese vessel to his fleet, bringing it to six ships in total.

Drake's fleet suffered depravations on the extended Atlantic crossing, which included the loss of two ships. Coasting southward along the east coast of South America, he eventually arrived at the Bay of San Julian where, half a century earlier, Magellan had put to death a group of mutineers. Drake's men discovered the sun-bleached remains still hanging on the Spanish gibbets. Ironically, Drake had to deal with his own mutineer, namely Thomas Doughty, the co-commander of the enterprise. On 3 June 1578, Drake accused Doughty of witchcraft and mutiny, tried and executed him. After putting an end to this potential rebellion, Drake decided to over-winter in San Julian before attempting to round South America and enter the Pacific.

Before departure, Drake discovered that one more of his ships, the *Mary*, had rotting timbers, so she was scrapped and burned. A few weeks later the three remaining ships of his small fleet set sail for the Magellan Strait. Violent storms destroyed one of the remaining ships, the *Marigold*, while another, the *Elizabeth*, under the command of John Wynter, chose this moment to return to England, leaving only the *Pelican* to push on into the Pacific. While passing through the Magellan Strait, Drake and his men skirmished with the local people, possibly the first Europeans to encounter the indigenous tribes. At this point, Drake decided to rename his ship the *Golden Hind*, an altogether more impressive name. This was done in honour of his patron, Sir Christopher Hatton, whose crest was a golden hind, a female deer.

Drake moved slowly north along the west coast of South America, attacking and pillaging ports and towns as he went. He captured a few small Spanish ships, gaining intelligence from their charts, which were much more up-to-date than

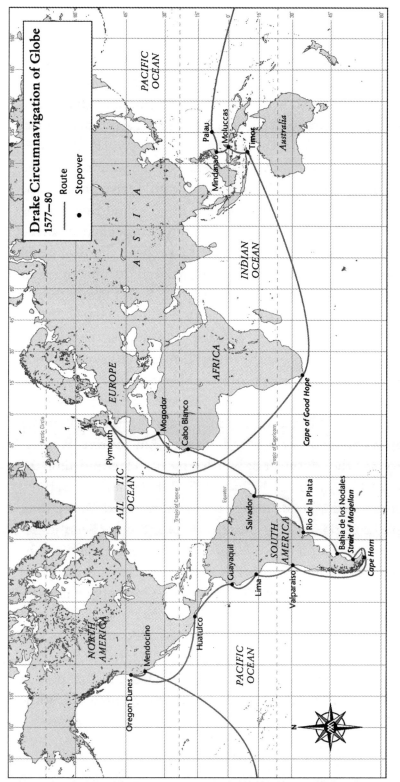

Map 28. Drake's voyage, which covered some 36,000 miles and took 1,020 days, was shrouded in secrecy; its real purpose was to pillage Spanish trade.

his own. He also succeeded in sacking the most important port of Valparaiso, where he captured a local vessel full of wine, which did much to cheer up his crew. Cruising further north, on the approaches to Lima, in Peru, Drake captured a Spanish ship laden with gold and gold pesos to the value of 37,000 ducats. Drake also gained intelligence on the whereabouts of a great galleon sailing to Manilla – this would turn out to be the *Nuestra Señiora de la Concepción*, which he would later call the *Cacafuego*. Drake eventually captured this vessel and it proved to be his most profitable capture. Among the chests of plates and jewellery were 80 pounds of gold and 26 tons of silver. After the capture of the *Cacafuego*, Drake sailed north, forever on the lookout for any other Spanish treasure ships.

Drake is reputed to have sailed as far as the 38th parallel, but here he turned back south, citing 'extreme and nipping cold' and the 'most vile, thick and stinking fogges'. He was also required under secret instructions to look out for the western entrance of the fabled 'water route' across America. (This was a possible route between islands or some kind of river connection across the continent of North America.) He landed on the coast of California on 17 June 1579 where he had carefully reconnoitered a good sheltered port, now called Drakes Bay, where he could repair his vessels. He maintained friendly relations with the local Miwok Indians, though during his stay, he claimed the land for the English Crown, calling it 'Nova Albion'. It's a claim that, 200 years later, even after the independence of the British Colonies and the American War of Independence, has technically never been refuted.

Drake left California, New Albion, on 23 July 1579; heading across the Pacific, he arrived in the Moluccas, also known as the Spice Islands, part of Portugal's sphere of influence. While in the islands, the *Golden Hind* was almost lost when it ran aground on a hidden reef. The crew offloaded part of her cargo and three days later, with favourable tides and winds behind them, they refloated their ship. Drake also became involved in local affairs for a while before sailing on across the Indian Ocean, rounding the Cape of Good Hope and reaching Sierra Leone, a region that he had visited previously, on 22 July 1580.

The weather-beaten *Golden Hind* finally arrived in Plymouth on 26 September 1580; the queen's half share of her cargo was worth more than all her other revenues combined for that year.

Queen Elizabeth I decided to keep the details of Drake's voyage a state secret on pain of death, though he was given a knighthood. He was the first Englishman to circumnavigate the world and begin to understand the shape of the earth's continents. Like any good seafarer, he was able to sketch a headland or bay as a location for an anchorage or a familiar landing place, but almost all maps showing his voyages were created by others, based on the accounts he published. His world route (1577–80) was published in a map by Nicola van Sype in 1581.

SIR WALTER RALEIGH

Sir Walter Raleigh, born in southwest England on 22 January 1552 to a well-connected and wealthy family, became an explorer, soldier, politician, investor, spy, landed gentleman and, in his spare time, a writer and poet.

In 1569 Raleigh served the Protestant Huguenot cause in the French religious wars and witnessed the Battle of Moncontour in the same year. His movements over the next few years are a little confused; in 1572 he spent a year at Oxford but left without a degree. He then studied law in London, which again was not completed, then returned to France, a restless soul. In about 1576, he returned to England and by 1579 he was in Ireland, his mission to take part in the suppression of the Desmond Rebellions (revolts by feudal lords in southwest Ireland against the expanding power of the English Crown). He was successful in this role, receiving some 40,000 acres of confiscated land, making him one of the largest landowners in the province of Munster. On this great estate, he tried to encourage Protestant settlement, though in this endeavour he did not succeed.

Nevertheless, with his Irish estates and family connections, Raleigh moved in the right circles and, in 1584, he was granted a Royal Charter, which granted him the right to explore, colonise and establish rule over any 'lands and territories not actually possessed by any Christian prince'. The Royal Charter covered a period of seven years in which Raleigh and his friends could attempt to establish a viable colony; it also allowed his backers to keep 80 per cent of all gold and silver found, with the balance going to the Crown. This seemed a pretty good deal – after all, the Spanish were doing well, and their ships loaded with goods and treasure were streaming back to Spain from their possessions in the New World.

In 1584 Raleigh dispatched two ships commanded by Philip Amadas and Arthur Barlowe (Raleigh himself would never clap eyes on any part of North America), which entered an inlet on the Outer Banks of what is now North Carolina (the local name for the place was Hatarask).

The explorers established good relations with the local people and two of them, Manteo and Wanchese, sailed back to

England with Amadas and Barlowe. Raleigh, attending the court of Elizabeth, was much encouraged by the reports he received and named the area Virginia to attract the Virgin Queen's support. In 1585, a full colonising expedition was sent under the command of its governor Ralph Lane with 107 settlers on board. They landed in August 1585 on the northern end of Roanoke Island and there, with the permission of the local chief, they built a fort. As it was late in the season, it was just as well they had a reasonable relationship with the locals, who provided them with maize and other supplies. The colonists explored the region around Roanoke, reaching Chesapeake Bay to the north and travelling inland along the Chowan and Pamlico rivers. However, arguments and petty squabbles became endemic in the small community and relations with the Indians soon deteriorated. In June 1586, Francis Drake's fleet called into Roanoke, no doubt to check the location of a possible base from which to raid Spanish trade. He found a demoralised community who immediately accepted the offer of a lift home to England.

A new expedition led by John White carrying over 100 settlers was sent in 1587. This group included Manteo, the Indian who had voyaged from America to England, who was made 'Lord of Roanoke'. White then sailed back to England in order to send supplies and recruit further settlers to the colony. The ineptitude that had marked previous ventures was now compounded by the arrival of the Spanish Armada off the south coast of England, and White was unable to send the planned supplies. White did not return until 18 August 1590, when he found the settlement abandoned. There was no sign of a struggle and many buildings had been taken down, indicating that the move had taken place over a period of

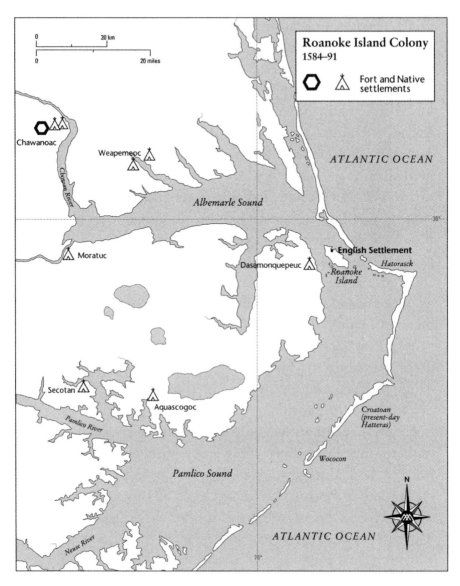

Map 29. *This colony was England's first attempt to found a permanent settlement in North America in 1585.*

time. There was no sign of the 118 men, women and children who had been left behind. The only clues were the word 'CROATOAN' carved on a fencepost and the letters 'C-R-O' carved on a tree at the edge of the settlement. The fate of the colonists remains a mystery and no conclusive explanation has yet been found. Two further expeditions were eventually sent in 1602 and 1603 and both failed to establish viable settlements.

In 1594 Raleigh acquired a Spanish account of a legendary city of gold located somewhere on the upper reaches of the Caroní River. He set off in 1595 to explore the coastline of Guiana, now Guyana, and along the coast of Venezuela. In 1596 his discoveries were published in his work *The Discovery of Guiana*, and its exaggerated claims did much to contribute to the creation of the El Dorado legend.

From 1596 to 1603 Raleigh took part in a number of operations, including the capture of Cádiz, and also held government appointments. However, when the queen died on 24 March 1603, the favour upon which his political existence relied disappeared with her. James I, the new monarch, was less than warmly disposed toward Raleigh, who was arrested on 19 July 1603 on charges of treason and imprisoned in the Tower of London in a rather well-fitted cell in the Bloody Tower. He was tried and found guilty, but the new king spared his life. Raleigh remained in some comfort in the Tower, and here he wrote the first part of his *History of the World* and managed to beget a son.

In 1617 he was pardoned and granted permission to undertake a second expedition to Guiana and Venezuela, looking for El Dorado. However, at the mouth of the Orinoco River a group of his men attacked a Spanish outpost in direct violation of Raleigh's pardon and parole. When Raleigh returned to England an outraged Spanish ambassador demanded the reinstatement of his death sentence and the king had little choice but to agree.

Sir Walter Raleigh was beheaded at the Palace of Westminster on 29 October 1619. After inspecting the executioner's axe, his last words were, 'Strike, man, strike.'

9

MERCATOR NAVIGATES THE WORLD

All world maps are a compromise in some way or other, and Mercator's solution came at a time when explorers and navigators were still involved in revealing large parts of our world that were little understood or not known at all.

Geert de Kremer was born on 5 March 1512 in the east Flemish town of Rupelmonde, then part of the Spanish or Habsburg Netherlands, which in turn was part of the Holy Roman Empire, and is now simply in Belgium. He is better known to us as Gerardus Mercator, and his 'Mercator' map projection is probably the best-known map projection in history.

After the death of Geert's father, Hubert, the young 14-year-old was taken under the guardianship of his uncle, Gisbert de Kremer. Gisbert, who was a priest of some standing in the local area and hoped young Geert would follow in his footsteps, sent his nephew to the school of the Brethren of the Common Life in 's-Hertogenbosch. However, alongside Geert's Bible studies, he was encouraged by the school's influential headmaster, Georgius Macropedius, to also study the philosophy of Aristotle, the natural history of Pliny and, perhaps most importantly, the geography of Ptolemy. Geert was to have some difficulty reconciling church dogma with classical learning, and this would trouble him in later years. Geert could read, write and speak in Latin – it was the language of the school's lessons – and he even gave himself a new name by 'Latinising' the meaning of his family name, becoming Gerardus Mercator Rupelmundanus. He was now ready for university.

Mercator entered the University of Leuven in 1530, and was entered on the university roll using his full Latinised name. He was also entered as a pauper, although his uncle Gisbert was probably sending him a little pocket money every now and then. Of the students that he rubbed shoulders with, many came from influential families and would one day make their mark. One of these students was Antoine Perrenot de Granvelle, who would rise to become a Burgundian statesman, a cardinal and the minister of the Spanish Habsburg state; he would become a lifelong friend of Mercator. There were many others like Antoine in the university. In 1534 Mercator achieved his degree, as a *Magister*, which is similar to a doctorate in modern terms. It covered philosophy and theology, following the principle of scholasticism, which was a method of critical thought used in teaching by academics in medieval universities.

In the normal run of things, having become a *Magister*, Mercator would have carried on with his studies, but he was troubled by the contradictions he saw between biblical studies and his own observations. This was, of course, at a time when Martin Luther challenged the dogma and approach of the Church. Mercator and his university friends no doubt discussed this philosophical clash between religion and science. If any of their discussions had come to the notice of the university authorities, they would have been considered heresy. It seems that the 20-year-old was careful enough not to put anything in writing. He decided to leave the university in 1534 and move to Antwerp, where he devoted his time to the contemplation of philosophy, though the contradiction between biblical teaching and his increasing knowledge of the world of geography would trouble him for the rest of his life.

Mercator remained in touch with other mathematicians and geographers from his student days; among them was Franciscus Monachus, born Frans Smunck, a Franciscan monk who was also the product of Leuven University and a somewhat controversial figure, who suffered from similar religious doubts to Mercator. Monarchus's own views on geography were based entirely on his own investigations, observations and research, which must have influenced the young Mercator.

While in Antwerp, Mercator apprenticed himself to Gaspar van der Heyden, a gifted goldsmith, engraver and maker of precision astronomical instruments and terrestrial and celestial globes. In the early 1530s, Van der Heyden created his first terrestrial globe in collaboration with Monachus, and Mercator contributed the typography. At the time, Mercator was working in Van der Heyden's workshop where a second globe was soon produced with the assistance of Gemma Frisius, a physician, mathematician and cartographer, with a new one planned in 1535, which was completed in 1536, showing the latest geographical discoveries. All this our hero, Mercator, would have witnessed. It may have been at this point that he put aside his theological problems to concentrate on geography. In later life, he wrote:

> Since my youth, geography has been for me the primary subject of study. I like not only the description of the earth but the structure of the whole machine of the world.
>
> Ptolemy's *Geography*, Introduction, 1578

At the end of 1534, the 22-year-old Mercator went back to university, applying himself to the study of geography, mathematics and astronomy under the guidance of Gemma Frisius. Gradually Mercator mastered all the elements of mathematics and astronomy that he would need. The university then granted him the right to teach private students, and while he was thus employed, he also mastered the skills of working in brass and engraving. He was closely involved in the creation of Van der Heyden's 1536 globe, upon which Mercator engraved the text, including his own name – the first time it was seen in public. In 1538 Mercator produced his first map of the world, usually referred to as the *Orbis imago*.

Now 30 years old, Mercator was beginning to feel increasingly confident about his abilities and his future. He had published his maps and was now in correspondence with influential people throughout Europe. However, politics intervened when the Duke of Cleves set about exploiting religious unrest in the Low Countries to his own ends. He besieged Leuven with his pro-Lutheran forces, and although the siege was eventually lifted, the damage and financial loss to its traders and merchants, including Mercator, was considerable. The second, potentially lethal, event was a call from the Inquisition.

Mercator, unlike most scholars of his age, did not travel very far from his home. Instead, he carried on prodigious correspondence, usually on the subject of geography, mathematics and philosophy. An example of new cartography arriving in the post would be samples of the work of Piri Reis, a naval officer and cartographer in the service of the Ottoman Empire, and particularly his book *Kitab-i Bahriye* ('Book of the Sea'), which contained many nautical instructions.

Mercator experienced a setback at the hands of the Inquisition. At no time during his correspondence did Mercator claim to be a Lutheran, but there are hints here and there that he had sympathies towards the Lutheran position. Consequently, in 1543, there came a knock at his door and he was arrested and taken by the Inquisition to Rupelmonde Castle, where he was cross-examined for six, or perhaps, seven months. Pressure from his many friends helped his situation and eventually he was released through lack of evidence.

Mercator was fortunate that his arrest and imprisonment did not seem to have any adverse effect on his customers. Typical of his correspondence during this period were the letters that passed between him and the English mathematician and astrologer John Dee. They had met after Dee graduated from Cambridge and 'went beyond the seas to speak and confer with some learned men' at Leuven University, and they remained friends for the rest of their lives. Dee received globes and instruments and in return sent the latest maps covering English explorations.

In 1552 Mercator moved from Leuven to Duisburg in the Duchy of Cleves, setting up a cartographic workshop there, which received the royal seal of approval from Emperor Charles V. He never stated why he moved, but the religious tolerance

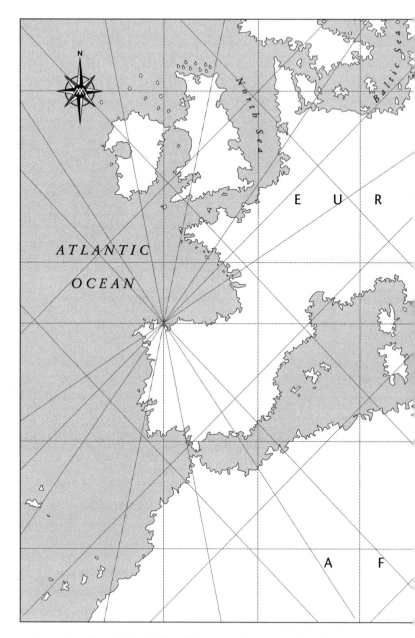

Map 30. Ahmed Muhiddin Piri, better known as Piri Reis, was an admiral in the Ottoman navy, and a geographer, navigator and cartographer of great skill.

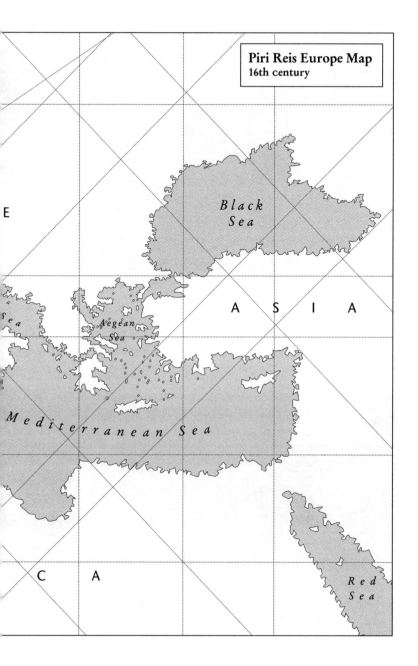

Piri Reis Europe Map
16th century

E

*Black
Sea*

A   S   I   A

S e a

*Aegean
Sea*

M e d i t e r r a n e a n   S e a

C   A

*R e d
S e a*

Map 31. *Gerardus Mercator created a map projection beloved by all navigators, because of its unique property of representing any course on a constant true bearing as a straight line.*

of the duchy and its quiet stability must have been attractive. Duke Wilhelm welcomed him and appointed him court cosmographer, and here he met the social elite, including the mayor, Walter Ghim, who would become Mercator's biographer. Over the following 17 years, Mercator conducted his business with a varied array of customers. His sons Arnold, Bartholomeus and Rumold joined the business and all made their own contributions.

The year 1569 was when Mercator published his most famous map, *Nova et aucta orbis terrae descriptio ad usum navigantium emendate accommodata* ('New and more complete representation of the terrestrial globe properly adapted for use in navigation'). Up to this date, most of Europe's mapmakers and explorers had based their work on elliptical projections that were derived from Ptolemy's latitude and longitude grid, which showed each degree of latitude or longitude as the same size. This meant that a straight-line compass bearing drawn on the map by a navigator – a rhumb line – would curve and have to be recalculated as they moved. Now, however, Mercator realised that a way to keep the navigator's rhumb lines straight so that they did not require constant recalculation would be to make the lines of latitude move away from each other as they moved north or south of the equator, thus ensuring that a 90-degree angle would be preserved between the lines of latitude and longitude. This map gives you an idea of how Mercator's projection works; the actual mathematical formula is extremely complex and involves far too many headaches to explain here, but it can be found on the Internet if you are a sucker for punishment and have a couple of days to spare. (It also helps if you have a loving family to help you recover.)

The projection is a navigator's delight and is still used by chartmakers today. The main criticism of it is obvious, though – as you move the world from a sphere to a cylinder, then unroll the cylinder to the flat, with latitude and longitude as straight lines, you will notice the equator is the only latitude that gives correct distance. As you move north or south the distortion increases – thus Greenland, for example, is bigger than South America. Another criticism is a political one; in exaggerating the size of northern Europe it made that continent geographically dominant – at a time when Europe owned colonial empires.

Mercator continued his work in Duisburg, publishing an atlas of Europe in 1570–72, Ptolemy's *Geography* in 1578 and an atlas of 51 maps in 1585. In 1586, his wife of many years died, followed by his eldest son and business partner Arnold in 1587. During this period, Mercator wrote a number of philosophical and theological works but somehow still found time in 1589 to get married again, aged 77, to a local lady named Gertrude. He also arranged for her daughter to marry his youngest son Rumold.

In 1589, an atlas of 22 maps dedicated to Ferdinando I de' Medici, Grand Duke of Tuscany, was published, but in 1590 Mercator suffered what seems to have been a severe stroke, which left him incapacitated. He worked on with difficulty and completed a set of maps he had in hand but finally died after two more strokes in 1594. In 1595, his son Rumold published posthumously *Atlas, or Cosmographical Meditations Upon the Creation of the Universe, and the Universe as Created*, which was the first time the word 'atlas' was used as the title of a collection of maps. This first, posthumously published, 'atlas' was dedicated to Queen Elizabeth I of England.

10

TERRA AUSTRALIS ('SOUTH LAND')

Unlike other parts of the world explored by Europeans, Australia was a unique experience. So far, European exploration had been led by the idea of trade with organised or large tribal societies that could be exploited in some way or other, either by enslaving part of the population, by seizing valuable minerals and spices, or – of course – by territorial acquisition and colonisation.

The Australian experience was uniquely different; here, Europeans found that small tribes had established themselves across the vast continent-sized island and adapted perfectly to all the climatic environments that it had to offer, forming family groups – clans – then 'nations'. Research shows that during the pre-European period there were about 250 nations that were bound together by common traditions and languages, of which about 50 survive today, although to some degree, all the nations have a shared culture that is said to be around 40,000 years old but is probably a lot older. They maintained the oldest continuous art style on earth.

How these people got to Australia is still an unsolved mystery. There is no evidence of sea-going craft anywhere in

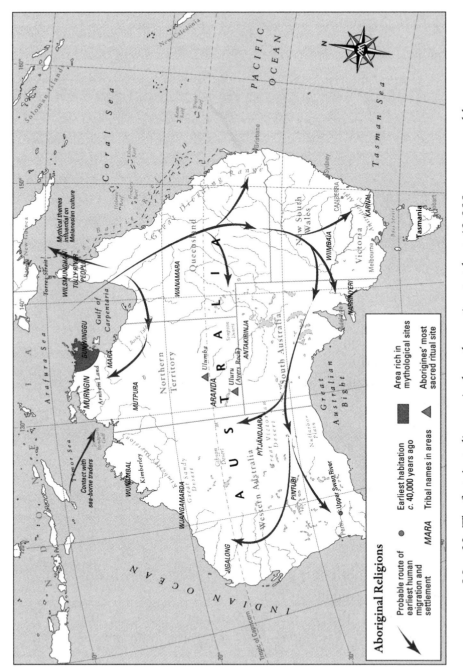

Aboriginal Religions

↗ (arrow)	Probable route of earliest human migration and settlement
●	Earliest habitation c. 40,000 years ago
MARA	Tribal names in areas
▓ (shaded area)	Area rich in mythological sites
▲	Aborigines' most sacred ritual site

Map 32. The first Australians arrived on the continent at least 40,000 years ago and have retained the longest consistent art style recorded among all of the planet's peoples.

Aboriginal culture. Until quite recently, it was believed that the Aboriginal people had been in possession of Australia for only around four or five hundred years, until the carbon dating of ancient human remains proved Aboriginal possession over many millennia. Most conjectural dates for the arrival of humans in Australia are between 40,000–80,000 years ago. The sea crossing must have been made at a time when the coastline was at a glacial stage, which meant that Australia included New Guinea and that Tasmania was also joined to the main land mass. Getting there still involved a sea passage of about 96.5 kilometres (60 miles), though. It's possible that a group of coastal canoes were blown off course and arrived together in the new land; biologically there would have needed to be about 25 humans in the group to reproduce successfully. And of course, they would have been hunter-gatherers, adept at exploiting local resources. Whoever they were originally, it was the emerging Aboriginal societies that were discovered by the Europeans in the 16th, 17th and 18th centuries.

At the time of discovery, the Aboriginal population was thought to be around 350,000, although recent calculations put the number up to 750,000. At 7,686,850 square kilometres (2,967,910 square miles) the place was, by most standards, empty. Even if we take the higher population figure to be correct, this would mean one person for every 10 square kilometres (approximately 4 square miles) or so – easy to miss in a sail-by. Cook noticed that many locals seemed disinterested in him and his strange large ship as he sailed by. The Portuguese also recorded this phenomenon when exploring up the Amazon River for the first time; as they passed sizeable settlements, some people stared, some ran away, but most seemed not to notice and carried on with their work, as though the image was too strange to register or rationalise in the mind.

In the 1780s, Captain Cook's view of the Aborigines was that they lived in a 'happy condition' and in tune with their environment and had thus seen no reason to develop in ways other societies had around the world. Apparently, their language had no words for 'yesterday' or 'tomorrow'. Their culture seemed to walk at a different pace to that of the Europeans with whom they would later come into contact. Perhaps the defining characteristic of the Aboriginal people was that they lived in Australia, and in tens of thousands of years, little changed in the look of the land. The European settlers, on the other hand, lived in Australia, dug foundations, constructed settlements, took out mortgages, built fences everywhere, introduced domestic sheep and invasive rabbit species, and, despite the land's huge extent, began to change everything.

*   *   *

The Portuguese were probably the first Europeans to see the Australian coast when they established a base just 650 kilometres (404 miles) to the north of it, in Timor, first arriving there in 1515. It seems improbable that these adventurous souls did not spend at least some of their time investigating the region, on the lookout for new trading opportunities. There are a number of maps that seem to substantiate this idea, but they could also be maps of the conjectural continent, dating back to ancient times (see Ptolemy, page 28), which was supposed to exist in order to balance the earth. In any case, it was just as likely that the Portuguese did not keep their paperwork in order and that any record of their encounter with Australia has been lost.

The first recorded landing by Europeans was in 1606 by a Dutch ship called *Duyfken* under the command of Willem Janszoon. He sailed from the island of Java, where the Dutch had a well-established base, and his instructions were to search out trade possibilities in the great land of New Guinea and other east and south lands. The task was a bit vague, but nevertheless off he went, reaching Bantam on the west coast of New Guinea. On 18 November 1605 he left Bantam, heading south across the eastern part of the Arafura Sea and into what is now the Gulf of Carpentaria. On 26 February 1606, a landing was made on the mouth of the Pennefather River on the western shore of the Cape York Peninsula, but the locals were not at all welcoming: 10 of Janszoon's men were killed on several landing expeditions while attempting to make friendly contact and to trade. He charted 320 kilometres (199 miles) of the local coastline, believing it to be an extension of the island of New Guinea and, confusingly, called the place Nieu Zealand; this name didn't stick, however, and was used elsewhere. With a less than warm reception and nothing to show on the trade ledger, Janszoon sailed away. He did visit Australia again on 31 July 1618, or rather bumped into it, at a point 22 degrees south, which he assumed to be an island some 35 kilometres (22 miles) long, but he saw nothing of interest and sailed on into history to enjoy a successful and prosperous life, until he popped his clogs in 1630.

ABEL JANSZOON TASMAN

In October 1606, a Spanish ship under the command of Luís Vaz de Torres sailed from distant Peru to Manila, its route taking it through the passage that is now called the Torres Strait, proving that Australia was separated from New Guinea – not that many people knew this for quite a while. Between 1616 and 1642 more Dutchmen appeared and, little by little, an understanding of the eastern coastline of the vast lands (now named on many charts as New Holland) was to some degree established, though no attempt to colonise this huge area was attempted. In August 1642, Abel Janszoon Tasman, in the service of the Dutch East India Company, set sail on a mission to explore the Southern Ocean for a suspected land mass that apparently contained a supply of gold, which had erroneously appeared on a map of the southern region. Tasman left Batavia, the capital of the Dutch East Indies (modern-day Jakarta), on 14 August 1642 and headed for Mauritius, where he arrived on 5 September. He stayed on the island for a while, which was in Dutch hands between 1598 and 1710, to repair and refit his ships, the *Heemskerck* and the *Zeehaen*. Tasman then set sail again, heading southeast, carried along by the winds we now call the 'Roaring Forties'. On 7 November, the crew, influenced by the increasingly cold conditions, prevailed on the ships' council and a decision was made to set a course further north in the hope of finding warmer weather. They expected to arrive somewhere in the Solomon Islands, but instead, on 24 November, they arrived off the west coast of Tasmania, slightly to the north of what is now Macquarie Harbour. Tasman named the new land Van Diemen's Land, after the Governor General of the Dutch East India Company, Antonio Van

Diemen. He then turned south, cruising around the south coast of Van Diemen's Land, before turning northeast, where he tried several times to enter Adventure Bay. The weather proved too inclement, however, and drove him further out to sea to a place he inventively called Storm Bay. Eventually, two days later, he found a safe anchorage in the Marion Bay area where, a day or so later, a landing was made and the Dutch flag was placed, symbolising formal possession of the land, which was declared on 3 December 1642. For a further two days, Tasman followed the coast until, once again, he was caught by the Roaring Forties and blown further eastwards. At this point, he gave up his exploration of Van Diemen's Land and sailed where the wind took him.

On 13 December, Tasman sighted the northwest tip of the South Island of New Zealand, which he named Staten Landt in honour of the States General of the Dutch Parliament. He wrote in his journal that he believed his discovery might possibly be joined to a land of the same name at the southern tip of South America. After sailing a little to the north and then eastwards for five days, he eventually dropped anchor near what is now called Golden Bay and immediately sent boats ashore to search for and collect fresh water. However, a double-hulled canoe filled with Maoris, who were obviously troubled by this uninvited invasion, attacked one of these boats and proceeded to club four of his men to death – a worrying turn of events. Tasman decided to leave the area, especially after he observed more canoes, 11 of which were swarming with warriors. When one of these approached his ship, the *Heemskerck*, a cannon was fired, killing a Maori on the nearest canoe. A later study shows that Tasman blundered into a valuable agricultural area, which the local Maoris were only too ready to defend, mostly against

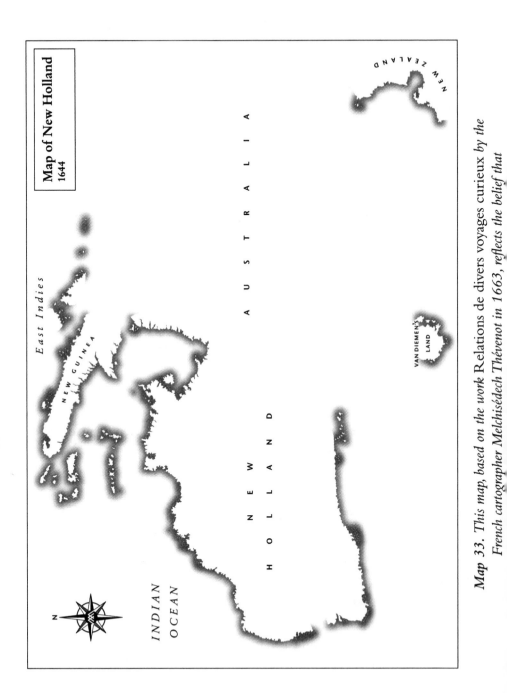

Map 33. This map, based on the work Relations de divers voyages curieux *by the French cartographer Melchisédech Thévenot in 1663, reflects the belief that*

other Maoris but also against anyone who posed a threat. Tasman sailed north, crossing the Cook Strait, which he mistook for a bight; not realising the south island was separated from the north island, he named it Zeehaen's Bight. Heading north, he eventually came across the Tongan archipelago on 20 January 1643, avoiding being wrecked in the Fiji Islands where numerous dangerous reefs were located. He continued around the northern coast of New Guinea, finally arriving back in Batavia on 15 June 1643.

Tasman did make a second voyage, leaving on 30 January 1644 with three ships. He sailed around the south coast of New Guinea in an attempt to find a passage leading to the eastern coast of New Holland. However, he managed to miss the 150 kilometres (93 miles) gap of the Torres Strait, no doubt put off by the numerous islands and reefs. He continued to map the shores of Carpentaria and the north coast of Australia, eventually arriving back in Batavia in August 1644. The Dutch East India Company was a little disappointed, to say the least, that he had failed to find the gold-bearing mystical land or to discover a useful trade route. It would be over a century before Europeans would again visit New Zealand and Tasmania.

WILLIAM DAMPIER

William Dampier, a son of the county of Somerset, was born in 1651 and first went to sea in around 1670 on merchant voyages. He witnessed both the naval battles of Schooneveld in the Franco-Dutch War; the English were helping the French at the time and he served in the combined fleet that attacked the Dutch – although the Dutch won both battles. For a number of years, Dampier suffered severe health problems, but he gradually recovered, married in 1679 and then left for the sea again.

Young William joined the crew of the buccaneer and scallywag Bartholomew Sharp – and I am sure he learnt much during this time. After years of pillage, Sharp was eventually brought to trial at the insistence of the Spanish. However, when standing before the High Court of the Admiralty, he produced and handed over a set of maps he had taken from a Spanish ship in 1681, which proved so useful to the Admiralty that a full pardon was received from the king.

After various ventures, Dampier signed on with privateer Charles Swan, captain of *The Cygnet*, and sailed with him across the Pacific in March 1686 in an attempt to seize a Spanish Manila galleon. In this they failed, but they made up for it by raiding the East Indies. After more nefarious activities, *The Cygnet* dropped anchor off the northwest coast of Australia near King Sound on 5 January 1688.

Dampier remained in Australia until 12 March while the ship underwent repairs, and during his stay he made notes and drawings of the people, flora and fauna he encountered there. By 1691, after further adventures, he had made his way back to England; he was almost penniless but in possession of his journals, and based on these he produced the book *New Voyage Round the World*, which became a popular sensation and attracted the beady eye of the Admiralty. They were so impressed that they gave Dampier a commission in the Royal Navy and a 26-gun vessel named HMS *Roebuck*, in which he was to explore the eastern coasts of New Holland. The expedition set off on 14 January 1699, too late in the season to sail round Cape Horn, so Dampier opted for the route around Africa, Cape Town and across the Indian Ocean. He arrived on the west coast of Australia on 6 August at a huge wide bay, its clear waters almost alive with sharks; inventively, he called the place Shark Bay.

When Dampier landed he began to set down in detail the first scientific survey of Australia's plants and animals, with drawings produced by his clerk, James Brand. Dampier followed the coastline in a northeasterly direction, recording what he found as he went. From the area of Lagrange Bay, he headed north to Timor, and after taking on supplies he headed east, rounded the northern coast of New Guinea and surveyed the islands to the east. His intention was to head south to explore the eastern coasts of New Holland, but, alas, his ship was in a serious state and just 100 miles from his destination he abandoned his plan and set a course for home. He managed to sail his increasingly leaky ship back across the Indian Ocean, around the Cape of Good Hope, and turned north into the Atlantic. His ship, on the point of sinking, ran aground on Ascension Island on 21 February 1701. The crew were marooned, though all were rescued, along with many charts, drawings and specimens, by a passing East Indiaman on 3 April. They arrived back in England in August 1701, and Dampier published his story of the expedition under the racy title *A Voyage to New Holland* in 1703.

On his return home, however, he was also court martialled for the death of his boatswain, John Norwood (for which he was acquitted), and the hard usage of his first lieutenant, George Fisher. The relationship between Dampier and Fisher had deteriorated to the point of cursing and violence, which had divided the crew of the *Roebuck*. When the ship had called into Bahia in Brazil, Dampier had had Fisher thrown into prison, and then set off, leaving him behind. Fisher eventually

*Map 34. The Europeans arrived in Australia in 1606
when a ship commanded by Willem Janszoon landed
at Cape York. Subsequently James Cook claimed
Australia for Britain.*

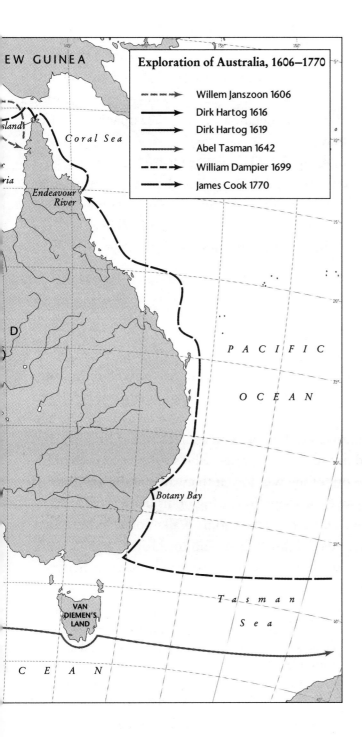

EW GUINEA

Coral Sea

Endeavour River

Exploration of Australia, 1606–1770

- - - → Willem Janszoon 1606
────→ Dirk Hartog 1616
───→ Dirk Hartog 1619
───→ Abel Tasman 1642
- - - → William Dampier 1699
─ ─ → James Cook 1770

P A C I F I C

O C E A N

Botany Bay

T a s m a n

VAN DIEMEN'S LAND

S e a

C E A N

made his way back to England, bearing a major grudge against his captain. Dampier was found guilty, lost his entire pay for the expedition and was deemed unfit to command a ship of His Majesty's Navy.

What could a chap do? Dampier went back to the life of a privateer and on one of his later voyages he rescued Alexander Selkirk, whose story is said to have inspired Daniel Defoe's *Robinson Crusoe*. He died in London in March 1715, £2,000 in debt – into each life a little rain must fall ...

JAMES COOK

James Cook, navigator, explorer, gifted cartographer and fellow of the Royal Society, was born on 7 November 1728 in Marton, Yorkshire*. After just five years of formal education he began work locally, helping his father, who was a farmer. By the age of 16 he had moved to Staiths as an apprentice haberdasher, but the work did not suit young James, who seems to have spent his time looking out of the window at the waves of the grey North Sea. His employer, perhaps despairing of his inattentive apprentice, introduced him to Captain John Walker of Whitby, a port just down the coast. John Walker and his brother Henry were well-known ship owners in the coal trade, plying between

---

*If you are in or visiting North Yorkshire and looking for 'Cook's Cottage' in the village of Great Ayton – Cook's parents' last home, regularly visited by the great explorer – you won't find it. In 1933 it was sold to Sir Wilfred Russell Grimwade, who had it shipped in 253 boxes and 40 barrels, including cuttings from the ivy that adorned the cottage, to be rebuilt in Fitzroy Gardens, Melbourne, Australia in 1934, the 100th anniversary of the foundation of the city. It is now a popular tourist destination.

northeast England and London. James was offered and accepted a merchant navy apprenticeship and his first ship was a collier (coal carrier), called the *Freelove*. On this and other ships in the coastal trade he applied himself to mathematics, astronomy and the finer arts of navigation. After his three-year apprenticeship, he worked in the Baltic trade and by 1755 was offered his first command, the *Friendship*. Just one month later, however, on 17 June 1755, he joined the Royal Navy.

Cook rose steadily through the ranks and in June 1757 passed his master's examination, qualifying him to handle a ship of the King's Navy. He joined the frigate HMS *Solebay* as sailing master on 30 June 1757 under Captain Robert Craig.

He served throughout the Seven Years' War in North America and during his service there, his skill at navigation and cartography was increasingly appreciated. He mapped the St Lawrence River, allowing General Wolfe to plan the attack on Quebec in 1759.

For five years, he mapped Newfoundland and adjacent coastlines, producing the first large-scale hydrographic surveys using triangulation* to establish accurate land outlines. Cook was capable and ambitious, stating that he intended to go not

---

*Triangulation is a method of surveying land that is based on a branch of mathematics called trigonometry, which deals with the properties of triangles. Triangulation begins with the measurement of a baseline and then sightings are taken from both ends of the baseline, aimed at a predetermined point in the distance. Then, by measuring the angles between the baseline and the sight lines to the distant point, a surveyor can fix the new point's exact position. That point then becomes one of the anchor points for a new baseline, from which another triangle is constructed. In this way, accurate lines of measurement can be extended for long distances in the form of chains of interlocking triangles. Triangulation is the principal method of making topographic maps.

only 'farther than any man has ever been before me, but as far as I think it possible for any man to go'.

Cook had his chance in 1768; on 26 May the Admiralty commissioned Cook to take command of a scientific expedition to the Pacific Ocean to observe the passage of Venus across the sun. It was hoped that the information he recorded, when added to observations taken at other locations around the globe, would establish the earth's distance from the sun. He was promoted to lieutenant in order to take command and the Royal Society also granted Cook 100 guineas (£110) in addition to his naval salary. At 39 years old, things were looking up.

On 26 August 1768 he set sail on HMS *Endeavour*, a sturdy vessel built in Whitby, Yorkshire, as the *Earl of Pembroke*, the kind of ship Cook was familiar with and well understood. Purchased by the Royal Navy and renamed in 1768 for this mission, she measured 98 feet long, 29 feet in the beam and weighed in at 366 tons, a home for a crew of 94 men. She would sail thousands of miles and last until 1778 when she was scuttled in Newport, Rhode Island, in the second year of the American Revolution. The *Endeavour* rounded Cape Horn, crossed the Indian Ocean and arrived off Tahiti in the Pacific Ocean on 13 April 1769.

The transit of Venus was observed and when that was done, Cook opened a packet of sealed orders prepared by the Admiralty, which contained detailed instructions to explore the region of the South Pacific and investigate the possible existence of a rich southern continent, postulated since the time of Ptolemy, *Terra Australis*. Cook sailed to New Zealand, where he mapped in detail the complete coastline of the main and adjacent smaller islands, with just a few minor errors. He then sailed westwards, reaching the southeastern coast of Australia on 19 April 1770. Four days later, the crew recorded the first sightings of human beings at Brush Island in what is now New South Wales. On 29 April the first landing was made at the place now known as Kurnell Peninsula, which Cook called Stingray Bay. A little later he crossed out that name, writing in 'Botany Bay' instead, perhaps inspired by the specimens collected by the expedition botanists Joseph Banks and Daniel Solander. (Joseph Banks went on to become president of the Royal Society, and was said to have brought 30,000 plant specimens back to England.) It was also during his stay that Cook recorded a meeting with a local tribe, which he knew (and I suppose they knew) as the Gweagal – the first Australian people recorded by this expedition.

Cook continued his voyage, heading north along Australia's eastern coast, landing to collect specimens on occasions. However, on 11 June HMS *Endeavour* ran aground on the Great Barrier Reef and, badly damaged, she coasted into the mouth of a nearby river (which Cook later named Endeavour River) in what is now Queensland, where it took seven weeks to make repairs. The voyage then resumed, reaching the tip of what Cook named Cape York, where he turned westwards. Seeking a vantage point from where he might be able to see a clear 'passage leading to the Indian Sea', he spotted a conical island, scaled the slope and signalled to his crew that he had found a navigable channel. Apparently loud cheering broke out on deck. According to his records, he took possession of the east coast in the name of the Crown and called this place Possession Island, although the Admiralty instructions did not specifically authorise this act. He sailed home via Dutch-controlled Batavia, the Cape of Good Hope and St Helena, arriving back in Britain in the port of Deal on 12 July 1771.

In August of that year, not long after his return, Cook was promoted to the rank of commander in recognition of his achievements, and was soon commissioned by the Admiralty on behalf of the Royal Society to head an expedition to verify the extent of the still hypothetical *Terra Australis*. The mission was to circumnavigate the globe as far to the south as possible, establishing the possible extent or even the existence of this landmass. The islands of New Zealand were proved not to be connected to a southern land mass and the eastern part of the continent-sized island of Australia also seemed unconnected.

Two ships, both stout Whitby-built vessels intended for the east coast trade of the kind very familiar to Cook, were purchased, refitted and renamed by the navy HMS *Resolution* and HMS *Adventure*. Cook commanded the former, Commander Tobias Furneaux the latter, and the ships sailed from Plymouth on 13 July 1772. They travelled south via Funchal in the Madeira Islands and the Cape Verde islands and arrived in Table Bay, South Africa, on 30 October with the crews in good health. On 17 January 1773, the ships passed the Antarctic Circle in the southern hemisphere's high summer, though as a precaution the crews had been issued with 'Fearnought' extra-warm jackets and trousers, supplied by the British Government in a rare moment of generosity. However, the ships encountered an area of thick Arctic fog on 8 February and became separated. Commander Furneaux navigated his ship towards the agreed meeting place, pre-arranged in case of just such an event, and on the way mapped the south and eastern coast of Van Diemen's Land, now known as Tasmania.*

---

* Here I declare an interest as my dear great-grandmother was born in what was then called Hobart Town, Tasmania, 97 years later. She was

He sailed on to the rendezvous at Queen Charlotte Sound, New Zealand, on 7 May 1773.

Meanwhile, Cook continued in a southeasterly direction, reaching 61 degrees 21 seconds south on 24 February. He then turned northeast, arriving in Dusky Sound on the southeastern tip of the South Island of New Zealand in mid-March. The ship took on fresh water and the crew rested, then sailed on to the agreed rendezvous, arriving 10 days after the HMS *Adventure* on 17 May.

The *Resolution* and *Adventure* sailed in consort from June to October, exploring the islands of the South Pacific. On a visit to Tahiti in mid-August, Mai, a local Pacific islander, boarded the *Adventure*, offering his local knowledge to the explorers. He would be the first Pacific islander to visit Britain and only the second in Europe. He would return to his home island of Ra'iatea, near Tahiti, on Cook's third voyage in 1776.

The ships continued their exploration, calling at the Friendly Islands and at Tonga before sailing on and returning to New Zealand. However, when the ships became separated during a storm on 22 October, they attempted to meet at the same rendezvous at Queen Charlotte Sound, but this time it did not work out. HMS *Resolution* arrived and waited until 26

---

the daughter of James Law, who was listed as a marble cutter. I surmise he arrived in Hobart Town as a guest of Her Majesty's Government – maybe he chipped off a piece of marble that was not his own. Despite transportation, his skill seems to have provided a good life in his new land, so much so that his daughter Helen was sent back to the old country to complete her education and be 'improved'. So, in her early eighties, she held me in her arms for a family photograph in Edinburgh; I remember nothing of the event, but I like to think I am a little improved by it by default.

November, when Cook, as agreed, left a message buried in a bottle describing his intention to continue his exploration of the South Atlantic, then return to New Zealand. Furneaux and the HMS *Adventure* arrived just four days later, on 30 November – he had been delayed after an encounter with a group of Maoris who had turned violent and killed a number of his men. He found Cook's message and left a reply, stating his intention to sail back to Britain. He departed Queen Charlotte Sound on 22 December 1773, sailing eastwards via Cape Horn and arriving back in Britain on 14 July 1774.

Cook, now enjoying summer in the southern latitudes, again headed south, sailing for 1,400 miles and reaching 71 degrees 10 seconds south at longitude 106 degrees 54 seconds west, where they encountered solid sea ice – this was the furthest point south that Cook achieved. Then he turned north to warmer climes and the crew thankfully removed their Fearnought winter warmers. Cook reached just below the equator, visiting the Friendly Islands, the Easter and Norfolk Islands, New Caledonia and Vanuatu, eventually returning to Queen Charlotte Sound, where he found Furneaux's reply. Cook again set sail eastwards over the wide empty Pacific on 10 November 1774 and on 17 December they sighted the western side of Tierra del Fuego and spent Christmas day at rest, naming the place Christmas Sound. The HMS *Resolution* then rounded Cape Horn and sailed into the South Atlantic, looking for a coastline that was believed to be part of a large southern continent, which later discoveries would reveal as Antarctica. They didn't find it, but Cook would find, name and claim South Georgia and the South Sandwich Islands. Cook was also able to predict that 'there is a tract of land near the Pole, which is the source of most of the ice which is spread over this vast Southern Ocean'.

The *Resolution* dropped anchor in Table Bay, Cape Town, on 21 May 1775 where she rested and refitted for five weeks. She eventually arrived home to England, to Spithead, on 30 July 1775, sailing via St Helena and Fernando de Noronha.

James Cook, now promoted to captain, was the toast of the scientific community and known well beyond. The House of Lords called him 'the first navigator of Europe' – he had come a long way from the farmer's cottage in Yorkshire. But he was by no means a spent force. The main objective in the third voyage commanded by the veteran explorer Captain James Cook was to discover the fabled water passage between the Atlantic and Pacific through or around North America: the Northwest Passage.

Originally the Admiralty planned to entrust this voyage to the experienced Royal Navy captain Charles Clerke; since Cook was in a sort of active retirement, he would act in a consultative role. Cook had studied all he could find on the North Pacific, including the accounts of Danish explorer Vitus Jonassen Bering's journeys in the region, and with this in mind, the Admiralty decided to place their faith in the veteran explorer instead – Captain James Cook would investigate the Pacific approach; again he would command HMS *Resolution*, while HMS *Discovery* would be commanded by Captain Charles Clerke. Meanwhile Lieutenant Richard Pickersgill would explore the Atlantic side, sailing into Hudson Bay in the frigate HMS *Lyon*. The planned search for the Northwest Passage was not openly discussed. Instead, the public learned of a voyage back to the South Pacific, which – among other objectives – would take the South Sea islander and society celebrity Mai, misnamed Omai, back to his home in the Pacific.

151

Captain James Cook set sail from Plymouth on 12 July 1776, while Captain Clerke delayed a little in London, sailing on 1 August. The ships met in Cape Town, both in a leaky condition, the *Resolution* arriving on 17 October and the *Discovery* following on 10 November. Both ships were re-caulked, made watertight and sailed together on 1 December – summer in the southern hemisphere.

Pushed eastwards by strong winds, they reached Van Diemen's Land on 26 January 1777. On their way across the Indian Ocean they had located, named and claimed the Prince Edward Islands and located the Kerguelen Islands, also called Desolation Islands, located in the Arctic latitudes – well named if you ask me. The two ships headed east, arriving at a favourite anchorage of Cook's, Queen Charlotte Sound, on 12 February. The local Maoris, on sighting the two ships, were in a state of some apprehension; four years earlier they had attacked and killed 10 men from HMS *Adventure* commanded by Furneaux. However, much to the Maoris' relief, Cook and his companions sailed two weeks later without exacting revenge.

The ships headed for Tahiti; however, easterly winds drove them towards the Cook Islands, which were sighted on 29 March. Cook had named these islands the Hervey Islands, but they were renamed the Cook Islands in 1827 by the German admiral Adam Johann von Krusenstern, who was in the service of Imperial Russia, and chose to honour the great explorer. They sailed on until they reached the Friendly Islands and stopped briefly at Palmerston Island, which Cook had also discovered in 1774, where they stayed from April until July, when they set out again for Tahiti and arrived there on 12 August.

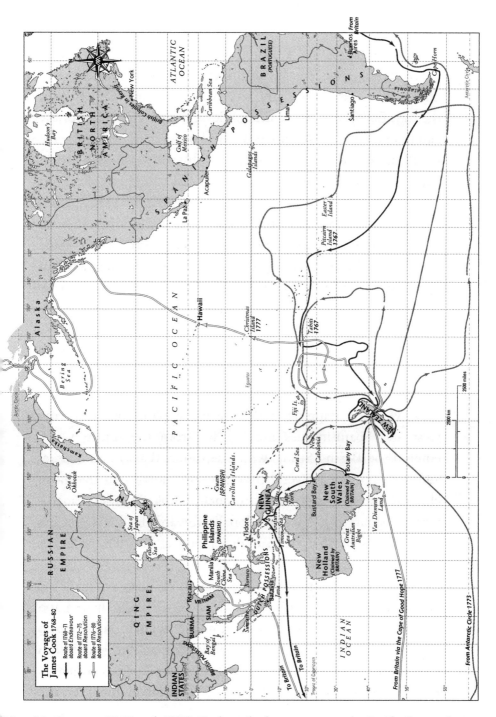

Map 35. Between 1768 and 1779 Cook made three voyages to the Pacific Ocean, during which he mapped the eastern coastline of Australia and the Hawaiian Islands.

After landing Omai at his home (matching the cover story), Cook and his ships left in December, after some delay, heading north until they eventually arrived in the Hawaiian Islands, the first Europeans to see them. Cook named these islands the Sandwich Islands after his boss the Fourth Earl of Sandwich, First Lord of the Admiralty – thankfully the name didn't stick. Leaving Hawaii, Cook headed northeast to explore the west coast of North America to the north of the known Spanish settlements. He arrived on the coast at Cape Foulweather, in modern Oregon; after being blown south a little he resumed his trek northwards, passing the Strait of Juan de Fuca without noticing it … most unusual.

The ships entered Nootka Sound, Vancouver Island, where they anchored near the village of Yuquot – the area is now called Resolution Cove. Here they traded with the locals until 26 April 1778 when they raised anchor and left to continue northwards, exploring the coastline and inlets and islands as they went, overlapping with Russian explorations from the west and Spanish ones moving up from the south. Cook entered the Bering Strait, which separates Russia and Alaska, and the ships tried several times to push north, but the harsh climatic conditions proved too much. Cook, frustrated at this retreat, began to behave irrationally towards his crew – it has been suggested that he had developed a stomach problem, causing his bad tempers.

The ships turned south for the Aleutian Islands, entering the port of Unalaska on 2 October, and again Cook prepared and re-caulked his ships. While in port he met the Russian Gerasim Izmailov, a navigator and promoter of Russian settlement with interests in what was becoming known as Russian America. Izmailov was responsible for the first detailed maps of the

Aleutian Islands, but whether Cook got a glimpse of them is unrecorded. After a three-week stay, Cook set sail for the Sandwich Islands (now the Hawaiian Islands) on 24 October 1778, arriving off the island of Maui on 26 November.

The two ships cruised around the Sandwich Islands for around eight weeks, and who wouldn't after the frozen conditions of the Bering Strait? They were searching for a safe anchorage and finally set eyes upon Kealakekua Bay on the main island of Hawaii. Here they anchored on 17 January 1779. Shortly after arriving they met with the local people, particularly a chief called Palea, and Koa'a, a priest, who came aboard Cook's ship and then insisted that Cook should accompany them to a ceremony on the nearby shore. This turned out to be a rather lengthy and convoluted ceremony, which coincided with a harvest festival among the local people, called Makahiki, which involved the worship of the Polynesian god Lono. Here was a great coincidence; the white sails of Cook's ship HMS *Resolution* matched the traditional image of the god Lono, who was often associated with white banners and had promised to return to earth by sea, which led to Cook being linked with Lono. Cook came to be considered, at least by a number of locals, a reincarnation of Lono. I have always thought mapmakers and mapmaking something special, but this was probably going a bit far. This view of events was first presented by a crew member of the *Resolution* and has since been challenged.

On the departure of the expedition after a month's stay, while they were still in sight of land, the foremast of the *Resolution* split and broke, forcing an immediate return to Kealakekua Bay to make good the damage. The surprising return of the two ships was not only unexpected but unwelcome by the Hawaiians – the festival involving the god Lono was over and so it seems was Cook's association with this god. Out of curiosity the local people swarmed over the strange and exotic vessels and a certain amount of petty theft occurred. Cook managed by one stratagem or other to get anything valuable back, but these visits were now mixed with bad humour, quarrels and occasional violence. A ship's boat was taken; this was a vital part of the ship's capability and Cook was determined to get it back.

Cook tried to lure the local king Kalani'ōpu'u aboard his ship, intending to use him as a hostage against the boat's return. Kalani'ōpu'u was somehow forewarned and this led to a confrontation with the local people – more arrived on the scene, some of them armed. Cook and his men attempted to withdraw to the beach and the safety of the boats there, but in the hand-to-hand struggle, Cook was struck on the head by a club and fell into the surf, where he was finished off by spear thrusts. His body was dragged away by the enraged Hawaiians and so ended Cook's land journey.

As the Hawaiians' anger subsided, Cook's body was retained by the tribal leaders and underwent various rituals that would render the body to just its bones, which would then be used as religious icons. However, an appeal made by the crew of his ship must have struck a chord with the islanders, who then returned his body to the British. In line with Royal Navy practice, Cook's crew buried his remains at sea.

Captain Clerke took command, though he was in no great shape himself, being in the final stages of tuberculosis. He took the ships back to the North Pacific, stopping on the Kamchatka Peninsula in the Russian Far East where he took on supplies. He then made the expedition's final attempt to pass beyond the

Bering Strait, which failed, and so the ships turned back, stopping at Petropavlovsk where Captain Clerke died on 22 August 1779. A report was sent overland through the Russian Empire and European states, and finally arrived in London five months later.

HMS *Resolution* and HMS *Discovery* were now commanded by John Gore and James King respectively. They set sail for the long voyage home, passing down the western coast of Japan, along the coast of China to Macao, following the East India trade route across the Indian Ocean via Cape Town and north up the Atlantic. As they approached the British Isles a vicious gale blew them so far north that the first landfall they could make was at Stromness in the Orkney Islands, so the ships coasted their way southwards and eventually docked at Sheerness on 4 October 1780. With news of Cook's and Clerke's deaths already known via the report sent from far-off Petropavlovsk, it was something of a subdued welcome.

The Admiralty entrusted the editing of Cook's journals, and journals kept by other members of the expedition, to Dr John Douglas, who finally completed the work and published it in June 1784. It ran to three volumes, totalling 1,617 pages, with 87 plates including maps, and cost the scorching price of £4. 14*s*. 6*d*. It was a bestseller and the print run sold out in three days. If you are lucky, you can pick up a copy today for about £4,000, plus postage and packaging. Worth every penny if you ask me.

11

ENSLAVED

In the late seventies and early eighties my family holidays were spent in Cornwall, though there were occasional forays into Europe. I was keen for the children to be good Europeans; after all, the UK had recently joined the European Economic Community – about time we Europeans made a better job of it, I thought, having spent some time now plotting the continent's blood-soaked past.

But it seemed Cornwall was the place to recharge. The hotel we stayed at was set in a small hamlet on its north coast, and had a long garden leading down to the shore. On days not suitable for the beach we would sally forth to inspect Cornwall more closely, and on one particular day after breakfast, maps were consulted and we chose to head for St Ives, then across to the south coast at Marazion. We set off under low-hanging cloud, but as we came within sight of Mount's Bay the sky cleared and the sun shone; ice creams all round, then, after cleaning ourselves up, we decided on a boat trip – always a popular choice. This took us around St Michael's Mount, which, according to the sea-spattered leaflet I read as we puttered across and around the Mount, was originally given by

Edward the Confessor to the Norman Abbey of Mont-Saint-Michel, which held it until 1424.

History forgotten, we voyaged around the Mount and back to the shore, where it was now time for lunch and a discovery. Sea air is good for the appetite, so we found a hotel doing lunches with a nice view over the bay, and ordered. While awaiting delivery I wandered across the dining room to inspect a long row of prints, each with a descriptive plaque of 60 words or so. This display had been organised by the wife of the owner, a keen local historian named Jenny. Along the row, the subjects covered were smuggling and mining, but the last four pictures caught my eye. They showed Barbary corsairs raiding the Cornish coast on slaving expeditions in the early 17th century, and one caption claimed that 'over 200 [Cornish locals were] taken into captivity, men, women and children, [the] village [was] left empty except for [the] old and lame'. I later spoke to Jenny and when I asked her about the raids she said, 'Yes, they happened all right, and occurred over a period of about 200 years or so.' Where was the Navy? I asked. 'Fighting the Dutch or the French. We were exposed out here in the west. By the way, the corsairs raided Ireland as well.'

My head was spinning a bit; how had this bit of history escaped me? Now was the time for a little research. Back in the bar of our hotel, I soon got talking to a pair of local brothers, who had been fishermen but now followed safer livings in the tourist trade. I mentioned the story of the North Africa privateers, and they both knew the stories of the slavers – 'At that time they Moors 'ad real fast boats, see, much 'andier than ours,' they told me, and they described the kind of Moorish boats and ships at sea as though it had happened yesterday. 'The Cornish boats were family owned and manned, so when the crew was taken for slaves the family lost their means of living – no protection for us, even a couple of miles out.' They went on to explain that Lundy Island off the north coast of Devon was a nest of slavers and pirates and even for a period of five years flew the flag of the Ottoman Empire. Here was another part of the story I had never heard before and I was determined to find out more.

The next time I was in London I spent a day in the London Library. This was in the pre-Google days, so little by little I uncovered a story of raids on the coasts of Europe that lasted for hundreds of years. I decided to concentrate on the period from the fall of Granada, the last Moorish state in Spain, in 1492, to the final campaigns against the Barbary States by the European states and the young United States in 1815–16.

Piracy was as common around the coasts of Europe as any other busy sea lane around the world and was caused by all the usual motivations for maritime crime that have been present since classical times, and probably before. The Christian reconquest of Moorish Spain had been delayed by civil war, but after the union of Aragon and Castile in 1479 the last Muslim kingdom of Granada fell in 1492. That left a fault line through the waters of the Mediterranean Sea, from the Straits of Gibraltar to the Balkans. Almost all the sea's islands were still Christian north and west of the line, and Muslim to the south and east. Just to complicate this neat picture, after the fall of the great Christian city of Constantinople in 1453 to the expanding Muslim power of the Ottoman Empire, the empire began expanding into the Balkans, taking millions of Christian subjects within its realm.

In Muslim lands, the institution of slavery was widely practised, and it was a complex system that was an important part of the economy. As a slave you could still aspire to rise through the ranks, though it would mean some sacrifice, namely castration – for instance, if you were the guardian of the harem. With such incentives, you could easily lose interest in promotion. Constantinople (Istanbul) recorded that in 1609, 20 per cent of its population was made up of slaves, who came from the Caucasus region, Europe, Central Asia and as far south as sub-Saharan Africa.

Within the Ottoman Empire there was a special collection, sometimes called the 'blood tax', which entailed taking young Christian boys from the Balkans and Anatolia to be converted to Islam and trained for elite units in the army – the Janissaries. It was from this group of slave soldiers that many of the great commanders of the empire were drawn. I am happy to record that this particular group remained united with their testicles, even when achieving high rank – after all, they wanted happy soldiers in the empire. It should be said that this 'blood tax' was abolished in 1703 and slavery within the empire came to an end in 1924 with the establishment of the Turkish Republic. Slavery was abolished in Iran in 1929, and in Oman in 1970.

In Europe, the question of slavery was dealt with in a fairly chaotic way, for instance:

- In the Holy Roman Empire, slavery was condemned as a violation of man's likeness to God in 1220.
- In Poland, the Statutes of Casimir the Great, *c.*1350, emancipated all non-free people.
- In Portugal, King Sebastian banned the enslavement of Native Americans, but not of hostiles, in 1570.
- In Lithuania, the Third Statute abolished slavery in 1588.

- Russia banned the sale of Russian slaves to Muslims in 1649.
- Slavery was abolished in all French territories and possessions in 1791.
- The Congress of Vienna declared its opposition to slavery in 1815.
- The United Kingdom made slavery illegal throughout the British Empire in 1833.

So we can see that during the period we are interested in, there was a lively market in the trafficking of human beings around the Mediterranean, and the religious divide gave some a certain legitimacy, although profit was always the most pressing objective.

The power and influence of the Ottoman Empire in North Africa, notably over what is now Morocco, Algeria and Tunisia, was limited and to all intents and purposes these countries conducted their own foreign policy, of which slaving was a part. Their major targets were European maritime trade and coastal settlements as far north as the Netherlands and even Iceland. Along the Barbary Coast of North Africa there were well-established slave markets that supplied local and Ottoman needs. They had a long-established supply of slaves from sub-Saharan Africa, and adding the Atlantic coasts of Europe to their traditional Mediterranean hunting grounds was not going to be a problem with all the experience and expertise they had at their disposal. To this they could add a surprising number of renegade Europeans, who arrived in the pirate states looking for any 'trading' opportunity. There were Dutch, French, Italian, Greek and English cutthroats among the many who found their way to ports like Algiers. It was no accident that they could navigate the inshore rocks and shoals of

Europe's coasts, as they had local intelligence. These renegades converted to Islam, at least in name, when sailing under the flags of the Barbary States.

A Dutchman named Jan Janszoon van Haarlem was one of those who had, in the English parlance of the time, 'gone Turk'. Now calling himself Murat Reis the Younger, Grand Admiral of Sale, he led a group known as the Sale Rovers. They came from the Republic of Sale on the Atlantic coast of Morocco, which had been founded by a large group of Moriscos – Spanish-born Muslims expelled from Catholic Spain after the reconquest. He seized the island of Lundy in the Bristol Channel and held it for five years from 1627, preying on the busy sea lanes nearby. During this time, he flew the flag of the Ottoman Empire over the island. Later exploits included a raid on Iceland and the famous Sack of Baltimore, Ireland, in 1631, when he took 108 people captive. They were sold in the usual way in the slave markets of the Barbary Coast, though somehow or other two determined souls made it back to their homeland. Janszoon began his career as a privateer sailing under the Dutch flag but gravitated towards the Barbary Coast where there was less control and more opportunity. He was symbolic of the sort of ruthless adventurer engaged in this piti-less trade, and was free to roam and raid at will. The forces ranged against him were less organised, as each nationality attempted mostly to look after its own affairs. Several nations – Spain, France, the Republic of Genoa and the Venetian Republic – possessed strong fleets, but they rarely acted together. For much of the period we are considering, one or two of the European states were at war with each other. During the period 1789 to 1814, almost the entire continent was in some state of war.

So, between 1492 and 1815, what was Europe's population loss due to the activity of the Barbary pirates? We know that, in order to maintain slave numbers within the Ottoman Empire and its near neighbours, some 8,000 new captives per year were required and, given the fact that capturing slaves was not a gentle process, there would be a certain wastage among the victims caused by the violence of the raiding practice, shock and exposure. Therefore about 10,000 per annum would be needed to get 8,000 people to the slave market. So, over the period of 324 years, this would mean a total of 3,240,000 people being taken from Europe to the slave markets of North Africa and, by and large, ending up in various provinces of the Ottoman Empire.

In 1784, the United States became involved in the Barbary trade when one of its merchant ships was taken and ransoms were demanded for the release of its crew. In 1785, Thomas Jefferson and John Adams crossed the Atlantic to London where they met with Tripoli's ambassador, Sidi Haji Abdrahaman, to negotiate a deal for the release of the Americans held captive. In a letter written by Jefferson and Adams to the American statesman John Jay on 28 March 1786, they report that Jefferson asked the ambassador directly, 'What gives you and your country the right to take people from the high seas?' The ambassador replied that 'The right was founded on the laws of the Prophet and that it is written in the Koran that all nations that did not answer to their authority were sinners and that it was their right and duty to make war upon them wherever they could be found and to make slaves of all they could take as prisoners and, moreover, that every Muslim who dies in the struggle was guaranteed to go to paradise.' Jefferson and Adams were both of the opinion that the tribute should not be paid; however, the United States Navy was in no state for immediate deployment and so the first tribute was paid to Tripoli and most of the American captives were released. By 1800, tributes paid to the Barbary States were costing the United States in the region of $1 million per year, or about 10 per cent of their national budget, depending on which account you read. Shortly before Jefferson's inauguration as president, Congress passed legislation that included provision for six frigates, which represented a considerable increase in the offensive power of the United States. Jefferson became president early in 1801 and Yusuf Karamanli, Pasha of Tripoli, immediately demanded a new tribute of $225,000 from him, which he refused. Consequently, on 10 May 1801, Tripoli declared war on the United States. Over the next four years this did not go well for Tripoli, as the Americans, with the help of Sweden, which was already at war with the Barbary States for similar reasons, blockaded the Barbary ports. On their first international deployment, the United States captured the Libyan city of Derna, which was the first time the flag of the United States was raised in victory on foreign soil. The possession of the city gave the Americans a great negotiating advantage in securing the release of many more hostages and a successful end to the war in 1805.

However, the peace was not to last. The Barbary pirates reverted to their annoying habits and, by 1807, had begun taking American ships again. Unfortunately, as was usually case, the Europeans were disunited and at war with each other, while the United States was at war with Britain from 1812 to 1814 – the British burnt down Washington in August 1814, which was rather distracting for President Madison's administration. With the defeat of Napoleon in 1814 came the first opportunity for the Europeans and the Americans to take action against the Barbary States. The second Barbary War was initiated by the United States in June 1815. An American naval squadron, commanded by Stephen Decatur and deployed off the North African coast, almost immediately seized two important Algerian warships. By the final week of June the American squadron was anchored off the port of Algiers, where the negotiations with the Dey (ruler) began. They were concluded on 3 July 1815: it was agreed that the Americans would return the two ships, the American captives would be released and the Dey would pay $10,000 in compensation for damage to American interests. All further tributes by the United States would cease, while full shipping rights in the Mediterranean would be recognised.

Following the Treaty of Vienna in 1815, the practice of piracy in the Barbary States was forbidden. Britain followed this up by sending a powerful squadron of ships to the region to finally stop their piracy and release the Christian captives. The local Barbary leaders quickly agreed to do so; however, the Dey of Algiers offered more resistance and negotiations became difficult. The leader of the expedition, Edward Pellew, 1st Viscount Exmouth, considered he had completed the negotiations and decided to return to England, but shortly afterwards Algerian troops massacred some 200 captives, who were considered to be under British protection. It was obvious for all to see that Edward Pellew's negotiations had been a failure. He immediately returned with his squadron, now supported by six Dutch warships, to the port of Algiers. The negotiations held on 27 August 1816 were again a failure. It was then decided that the British fleet should bombard the port of Algiers, which lasted for nine hours, severely damaging most of the Dey's ships and fortifications. The Dey was offered the same terms that had previously been rejected, and, standing among the smoking ruins of his city, he now accepted them and the agreement was signed on 24 September 1816. Over 1,000 Christian slaves were freed, and the era of the Barbary pirates effectively ended. France occupied the coast of Algeria in 1830 and Tunis in 1881, while the Italians occupied Tripoli in 1911, ensuring European control of the Mediterranean Sea.

12

THE CASSINIS' CONCEPTIONS

The Cassini family was of Italian origin, but Giovanni Domenico Cassini (1625–1712) became a French citizen and changed his name to Jean Dominique Cassini. With his scientific and mathematical skills, he found work at the Paris Observatory and while working there he discovered the four moons of Saturn. He seems to have been a little obsessed with Saturn and, after further lengthy study, he gave his name to the separate elements of the rings that surround that exotic planet.

One day, Cassini was approached by the Minister of Finances under King Louis XIV no less, Jean-Baptiste Colbert, who wanted Cassini's help to prepare an accurate topographical map of the Kingdom of France. When Cassini looked into the problem of mapping a state the size of France, he quickly became concerned with the major cartographical issue of the day, which was the determining of longitude. This required that the surveyor or observer knew the exact time at two widely separated points. This was not an easy undertaking, given the varying, and sometimes unreliable, nature of even the very best clocks and timepieces of that era. After some consideration, Cassini turned back to the stars; he developed an extremely complex method of calculating longitude based on, of all things, observations of Jupiter's moons. This required the use of a flat surface and large telescopes; although it was almost impossible to do at sea, it could be done with a fair degree of accuracy on land. Surveyors were soon deployed, trained in Cassini's method, and quickly numerous points had been accurately determined based on a number of key places across France. An outline map was then very carefully drawn, encompassing these selected points and places. When that new outline was placed carefully over the existing and accepted map of France, it was quickly realised that the county's actual size was much smaller than previously understood. When King Louis XIV visited Cassini at the Paris Observatory and witnessed the reduction in the extent of his kingdom, he exclaimed, 'My dear sir, your work has cost me half my estate!' and flounced off.

At this time Cassini was also creating a huge world map on the floor of the Tower Room of the Paris Observatory, and each time reliable and verified information came his way, following the latitude and longitude observations sent in by his many correspondents, it was carefully recorded on this world map. This was a decisive moment in the long history of cartography, as it was the first time that accurate information, scientifically examined and systematically organised, was applied to a world map. Little by little, the now familiar forms began to take shape, correcting the mistakes and misconceptions that had been inherited from the Middle Ages.

*   *   *

Back to the state of France. The map that had so offended King Louis XIV, though much more accurate than those that had gone before, did not include the technique of triangulation in its creation (see page 145); the maps now to be undertaken by the Cassini family, however, would. This was an important innovation, in that it produced greater accuracy in the surveying of land, as triangulation's inherent characteristic was not to distort. This, together with the training of the teams deployed, drastically improved the accuracy of their work in the field, often under difficult circumstances. The teams who were directed to employ their instruments and establish the accurate locations of towns, roads and geological areas in the landscape were reasonably educated men. Their instructions were to make sure of these physical measurements, but also to give the correct names to their locations, which, given the substantial dialectic and linguistic variations across the huge territories of France, was not as easy as it sounds. They were also instructed to adopt a consistent style in the representation of roads, rivers, towns, villages, churches, canals and forests. As the cartographic data arrived in Paris, Cassini and his team would validate the information before it was allowed to be included in the final map. Cassini's first objective was to determine part of the meridian of longitude through the whole length of France. Once this was agreed and established, he could then use the technique of triangulation to complete the rest of the map by relating new points added to those already established on the meridian. The Cassini objectives were:

• To measure distance by triangulation, establishing the exact position of locations.

- To measure the Kingdom and to determine the innumerable number of boroughs, towns and villages scattered throughout its territory.
- To represent that which is unchangeable in the landscape.

Following Jean Dominique Cassini's death on 14 September 1712, his son Jacques Cassini (Cassini II) succeeded him at the Paris Observatory in 1712 and was now ably assisted by his son, César-François (Cassini III), and they worked on this element of the survey together. The initial framework of survey points was completed in 1740. The resulting maps, numbering 18 in all at a scale of 1:86,400, were considered more accurate than anything that had gone before.

The work continued in 1750 with a survey of roads, rivers, towns and other landmarks and was still incomplete when César-François (Cassini III) died in 1784. Jacques Dominique (Cassini IV) carried on with the work, and finally finished the great national atlas of France in 1791, at a point when France was convulsed by the French Revolution. The anti-monarchist revolutionaries almost immediately threw the Cassinis into prison, contaminated as they were by their close relationship with the French monarchy and the former ruling elite. During the Revolution, taxation records and other administrative documents went up in smoke, but fortunately the Cassinis' cartographic research and their mapping did not suffer the same fate.

Jacques languished in prison for nine long months, eventually being released, lucky to be alive. He decided, wisely in my view, to retire quietly to the countryside and there he worked on the topographical, mineralogical and statistical atlas of France, which he published, again quietly, in 1818.

The original concept of the topographical atlas was published in 1793 by the French Academy of Sciences, and it set a new standard for systematic cartography. The work influenced other countries around the world and paved the way for other national mapping projects. The Cassini family and the cartographic teams they created and trained were the predecessors of all future national mapping organisations. The coastlines, river systems and road networks plotted on the maps are so accurate they fit extremely well when overlaid by modern satellite images. Well done, boys!

13

THE PROBLEM
WITH EMPIRES

The whole question of empires is a fascinating one for any mapmaker working in the field of history. The island of Britain had already been part of three empires by the beginning of the 13th century, starting with the Roman Empire, in which the island was involved from 43 CE to 410 CE, when the Legions packed their bags and marched away – a sort of Brexit in reverse. The island was left to defend itself, which it did for a while, but the departure of a professional army was a difficult gap to fill. The Roman army recruited a lot of its men from Germanic tribes, many of whom had left with them, but not all. Some returned to their German homeland, among the Angles, Saxons and others. These people already understood the layout of Britain and the opportunities it could offer, and under population pressure themselves, they decided to become illegal immigrants and headed back to the shores of the east coast of Britain. Over time, they landed in sufficient numbers to change the language and culture of most of lowland Britain, which they began to name after themselves – Angleland, Englaland.

The story does not end there, however; the second empire to which Britain belonged was the Scandinavian or North Sea

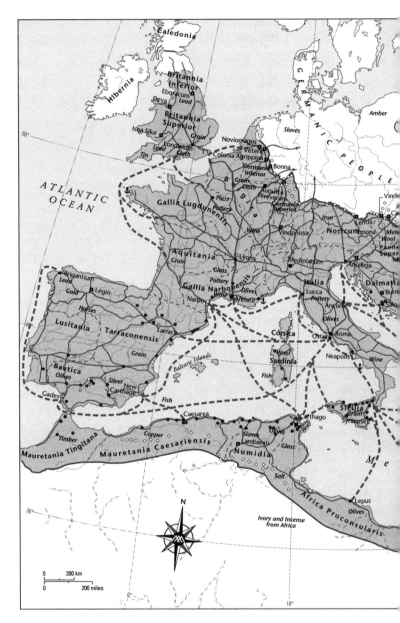

Map 36. By 180 CE, thanks to a plentiful supply of currency minted in the Roman Empire, a vigorous trade, mostly in agricultural products, was carried on throughout its territories.

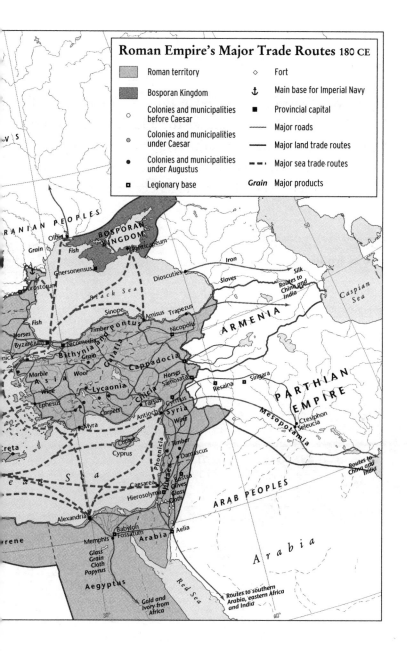

Roman Empire's Major Trade Routes 180 CE

▢ Roman territory	◇ Fort
▢ Bosporan Kingdom	⚓ Main base for Imperial Navy
○ Colonies and municipalities before Caesar	▪ Provincial capital
◉ Colonies and municipalities under Caesar	—— Major roads
● Colonies and municipalities under Augustus	—— Major land trade routes
	‒‒‐ Major sea trade routes
▣ Legionary base	*Grain* Major products

IRANIAN PEOPLES

BOSPORAN KINGDOM

Olbia
Grain
Fish
Panticapaeum
Chersonensus
Durostorum
Black Sea
Dioscutis
Slaves
Iron
Silk
Routes to China and India
Caspian Sea
Sinope
Fish
Amisus Trapezus
ARMENIA
Horses
Byzantium
Nicomedia
Pontus
Nicopolis
Timber
Bithynia
Galatia
Grain
Cappadocia
nica
Marble
Wool
Horses
Samosata
Singara
Asia
Wine
Lycaonia
Cilicia
Tarsus
Resaina
PARTHIAN EMPIRE
Ephesus
Carpets
Antioch
Syria
Mesopotamia
Ctesiphon
Seleucia
Myra
Wine
reta
Copper
Cyprus
Timber
Phoenicia
Damascus
Routes to China and India
Sea
e a n
Caesarea
Judea
Olives
ARAB PEOPLES
Hierosolyma
Glass
Cloth
Alexandria
Babylon
Fossatum
Aelia
Arabia
rene
Memphis
Arabia
Glass
Grain
Cloth
Papyrus
Aegyptus
Red Sea
Gold and Ivory from Africa
Routes to southern Arabia, eastern Africa and India

30°   40°   50°

Empire of King Knud (Canute) – though it did not last long, as he could not hold back the incoming tide either. This island had to go through another invasion, from the Normans, to eventually become part of the Angevin Empire of Henry II of England. This invasion had replaced the old English ruling class with a new French-speaking elite with connections across France and Flanders. New families now held power in the island and among them were the Plantagenets, originally from

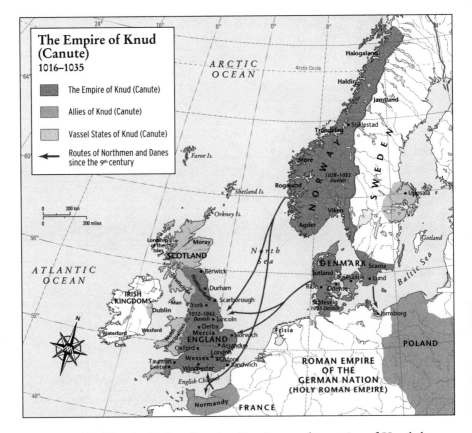

Map 37. The Anglo-Scandinavian Empire was the creation of Knud the Great, known in England as Canute, and existed between 1016 and 1035.

178

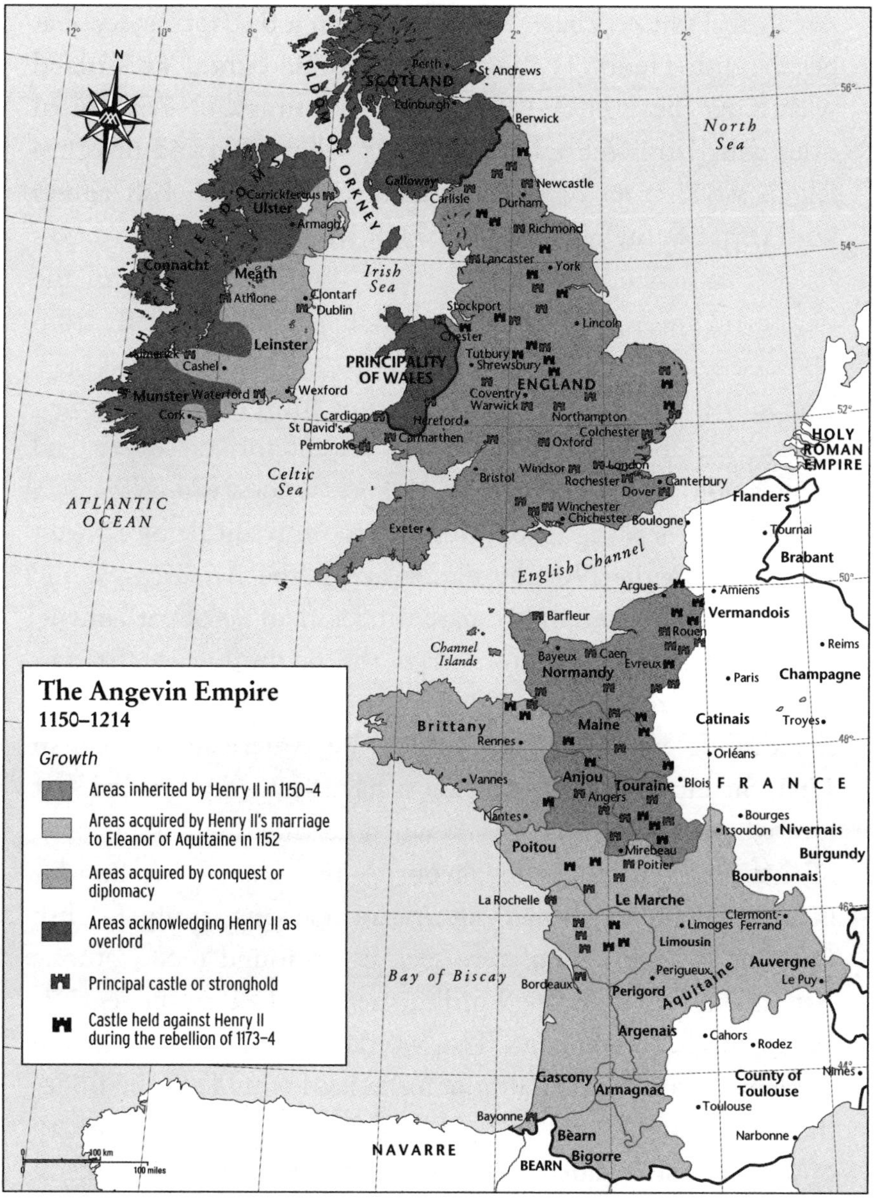

The Angevin Empire
1150–1214
Growth
Areas inherited by Henry II in 1150–4
Areas acquired by Henry II's marriage to Eleanor of Aquitaine in 1152
Areas acquired by conquest or diplomacy
Areas acknowledging Henry II as overlord
Principal castle or stronghold
Castle held against Henry II during the rebellion of 1173–4
SCOTLAND
EARLDOM OF ORKNEY
St Andrews
Edinburgh
Berwick
North Sea
Galloway
Carlisle
Newcastle
Durham
Richmond
Ulster
Carrickfergus
Armagh
Connacht
Meath
Athlone
Clontarf
Dublin
Irish Sea
Lancaster
York
Stockport
Lincoln
Chester
Leinster
Limerick
Cashel
PRINCIPALITY OF WALES
Tutbury
Shrewsbury
ENGLAND
Munster
Waterford
Wexford
Coventry
Warwick
Cork
Cardigan
Hereford
Northampton
St David's
Pembroke
Carmarthen
Colchester
Oxford
Celtic Sea
Windsor
London
Bristol
Rochester
Canterbury
Dover
HOLY ROMAN EMPIRE
Flanders
ATLANTIC OCEAN
Winchester
Chichester
Boulogne
Exeter
Tournai
Brabant
English Channel
Argues
Amiens
Vermandois
Barfleur
Rouen
Channel Islands
Bayeux
Caen
Evreux
Reims
Normandy
Paris
Champagne
Brittany
Maine
Catinais
Troyes
Rennes
Orléans
Anjou
Touraine
Blois
F R A N C E
Vannes
Angers
Bourges
Nantes
Issoudon
Nivernais
Poitou
Mirebeau
Poitier
Burgundy
Bourbonnais
La Rochelle
Le Marche
Clermont-Ferrand
Limoges
Limousin
Auvergne
Bay of Biscay
Perigueux
Le Puy
Bordeaux
Perigord
Aquitaine
Argenais
Cahors
Rodez
Gascony
County of Toulouse
Nîmes
Armagnac
Toulouse
Bayonne
Bèarn
Narbonne
NAVARRE
Bigorre
BEARN
100 km
100 miles

Map 38. The Angevin Empire was a collection of lands mostly in England and France, and is an early example of a composite state. It existed between 1154 and 1214.

Anjou in France. These 'Angevins' were a feudal dynasty that began with Henry II in 1154–89 and Richard I and ended with King John (1199–1216). The highest rank in their feudal titles was King of England; all their other titles and holdings fell in strict order behind England. (My English half agrees with this, but my Scottish half is not so sure.)

*　　*　　*

The fourth empire had existed in some form from 1583 when Queen Elizabeth I granted letters patent for 'discovery and exploration' to Sir Humphrey Gilbert, who was a soldier, an explorer, a leading supporter of the mythical Northwest Passage from the North Atlantic across the Americas to the riches of China, and in his spare time a member of parliament. This is regarded by historians as the beginning of the first British Empire, though I regard it as the fourth empire – in any case, it ended with the loss of the American colonies in 1783. Sir Humphrey Gilbert must have seemed a safe bet, but luck was not with him in 1583 when he sailed with a squadron of five ships, including his favourite, the *Squirrel*, heading for Newfoundland. He had significant backing from English Catholic investors who were seeking to found nice Catholic settlements based on the 9 million acres of land in the legend-ary region of Norembega. This would be handed out in lots to the generous investors, although the land would still be under the control of the Crown ... sort of. Alas for Gilbert, almost all of these investors faded away when faced with fines on account of their faith.

We must remind ourselves that this was after the union of English and Welsh law from 1535 onwards, so at best this was the founding of the English Empire, or perhaps the Anglo-Welsh Empire. We will come back to the other political unions that created the British Empire in 1707 and 1801 later.

Gilbert's ramshackle fleet, now numbering four ships (one had turned back, interestingly commanded by Walter Raleigh), was manned by all kinds of lawless misfits, the sweepings of England ports, but somehow it made it to Newfoundland. The fleet arrived in St John's and immediately faced a blockade organised by locals under the leadership of an outraged Englishman and port admiral, who had been roused by the piratical actions of one of Gilbert's commanders the previous year. This demonstration of outrage by an international community of fishermen and traders was eventually overcome and Gilbert was able to land, clutching his letters of patent from Queen Elizabeth I of England. She was set on having her piece of the New World, well away from the established Spanish territories. To take possession of Newfoundland, including lands 200 leagues to the north and south, Sir Humphrey Gilbert had to cut turf on the land in question, which symbolised the transfer of ownership of that land according to the common law of England. He accepted the gift of a little dog given to him by a crowd of locals, which he called Stella after the North Star. He then confirmed his authority over the township of St John's and all other fixtures and fittings and levied taxes on all enterprises and fisheries – I'm sure the population was delighted. 'The Queen deserves only the best!' cried the crowd in several languages, or maybe that's just what Gilbert thought they said.

Gilbert stayed in Newfoundland for a few weeks and, lacking supplies, made no attempt to found any new settlements. In any case, a large part of his crew was made up of freebooters and thieves – not the best material from which to build a new empire, he thought. Perhaps this unreliable labour force was the major flaw in his plan, but he also made a tactical error. The squadron left port to explore the locality, and he placed his largest ship, the *Delight*, in the lead; according to the other commanders, this was a mistake. She ran aground and sank, taking with her most of the crew and the bulk of the remaining supplies.

After this 'accident' Gilbert did consult his remaining commanders and on 31 August the squadron set a course for England. They made reasonable time until they came close to the Azores, where they encountered a fierce storm. Gilbert, in his favourite ship the *Squirrel*, had been warned that she was carrying too much topweight and too many guns, which was not good in bad weather, but he ignored all advice – at least he was consistent in this aspect of his character – and sailed on. His ship was almost overwhelmed by the storm on 9 September and, during the following night, when the *Golden Hind* came within hailing distance, Gilbert called out, 'We are as close to Heaven by sea as by land!' and was seen to be reading by a ship's lamp. Later that night the *Squirrel*'s lights disappeared and by early dawn she was nowhere to be seen; she had gone down with all hands. Incidentally, the book he was reputed to be reading was *Utopia* by Thomas More, published in 1516, a satirical work based on an island society located in a fictional new world.

Sir Humphrey Gilbert's final thought as he slipped below the waves may well have been framed by the religious and political thoughts of Thomas More, an ardent Catholic philosopher, which was a little dangerous in a Protestant kingdom.

That was probably the least of his worries as the final giant wave engulfed his ship. At least in his own way he had turned the first sod in founding an empire, the fourth in Britain's history.

* * *

In 1603, James VI of Scotland ascended to the English throne as James I of England and in the following year negotiated the ending of war with Spain; this became the Treaty of London, which ended 19 years of hostilities. Now, rather than preying on the revenues created by the already established Spanish Empire, England could focus on founding its own colonies. This was financed by the creation of joint-stock companies – for example, the East India Company chartered by Elizabeth I on 31 December 1600. Both the London Company and the Plymouth Company, established in 1606, covered the Americas.

During the period of hostilities with Spain there had been an earlier attempt to found a colony on the east coast of North America: Roanoke Island, founded in 1585 but found abandoned in 1590, due to a lack of consistent supply. And so it was the Caribbean that would provide a return on England's first colonial venture. There were several initial attempts to settle, all of which failed with surprising rapidity: for example, St Lucia in 1605 and Grenada in 1609. Eventually, however, with experience and practice, successful colonies were established in St Kitts in 1624 and Barbados in 1627 and elsewhere. With no discoveries of gold and silver on these islands, though, the colonies copied the system of sugar plantations pioneered by the Portuguese in Brazil, which was dependent on slave labour.

The English colonies would import indentured labour and use transported criminals as well, but it would be slavery that defined the plantation system over the islands of the Caribbean and in the later American colonies. Over the next two decades, the Caribbean became one of the most valued assets of the English Empire.

In 1607, England's first permanent settlement on the east coast of North America was founded, Jamestown, which was initially managed by the London Company, sometimes known as the Virginia Company of London. The lands granted to the company included the eastern coast from the 41st parallel in Long Island Sound south to the 34th parallel at Cape Fear; this was quite a chunk of land. The Jamestown settlement was established on the James River about 64 kilometres (40 miles) upstream from the river's mouth on Chesapeake Bay. In 1620, Plymouth was founded by Puritan settlers from England, then known as the Pilgrims, who were busy escaping from religious persecution. The Massachusetts Bay Colony was founded eight years later by the owners of the Massachusetts Bay Company and some 20,000 settlers headed for New England over the next 10 years or so. Failing to discover gold or other treasures, this new wave of settlers found their salvation in the production of crops for local consumption and, a little later, cash crops such as tobacco and indigo for export.

Many more English settlers, driven by similar ideas of religious freedom, followed these Pilgrims. Maryland, founded in 1634, was a new home for Roman Catholics and Rhode Island was founded the following year as a colony that tolerated, indeed welcomed, all religious persuasions. Connecticut,

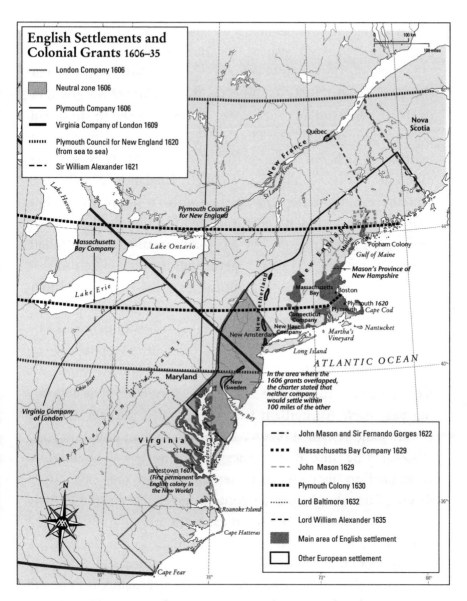

English Settlements and Colonial Grants 1606–35

- ━━━━ London Company 1606
- ▨ Neutral zone 1606
- ━━ Plymouth Company 1606
- ━━ Virginia Company of London 1609
- ┉┉┉ Plymouth Council for New England 1620 (from sea to sea)
- ─ ─ ─ Sir William Alexander 1621

Nova Scotia

Québec

New France

St Lawrence River

Lake Huron

Plymouth Council for New England

Massachusetts Bay Company

Lake Ontario

New England

Maine

Popham Colony
Gulf of Maine

Mason's Province of New Hampshire

Lake Erie

New Netherland

Massachusetts Bay

Boston

Plymouth 1620
Plymouth Cape Cod

Connecticut Company

New Haven Company

New Amsterdam

Martha's Vineyard

Nantucket

Long Island

ATLANTIC OCEAN

Ohio River

Maryland

New Sweden

In the area where the 1606 grants overlapped, the charter stated that neither company would settle within 100 miles of the other

Virginia Company of London

Appalachian Mountains

Virginia

St Mary's

Delaware Bay

Chesapeake Bay

Jamestown 1607
(First permanent English colony in the New World)

N

Roanoke Island

Cape Hatteras

Cape Fear

- ─ ─ · John Mason and Sir Fernando Gorges 1622
- ▪▪▪▪ Massachusetts Bay Company 1629
- ～～～ John Mason 1629
- ▪▪▪▪▪ Plymouth Colony 1630
- ┄┄┄ Lord Baltimore 1632
- ─ ─ ─ Lord William Alexander 1635
- ▨ Main area of English settlement
- ☐ Other European settlement

Map 39. Because of inaccurate maps and surveys, colonial grants to companies and influential individuals frequently overlapped.

founded in 1639, was decidedly in favour of Congregationalists and the province of Carolina was founded in 1663.

In 1651, as the colonies grew, it was decided by Parliament that only English ships would carry the product of the colonies to the motherland and vice versa. Previously Dutch ships had dominated this trade. The Dutch, of course, were very annoyed by this, which led to a series of Anglo-Dutch wars. The outcome of this would largely benefit England. In 1624, the Dutch created a settlement on the southern tip of Manhattan Island, which included a fort intended to protect other Dutch settlements in the region. The Manhattan settlement was called New Amsterdam and this, together with other Dutch settlements, would later be a political extension of the Dutch Republic.

On 27 August 1664, as a result of the Anglo-Dutch wars but during a period when England and the Dutch Republic were at peace, four English frigates sailed into New Amsterdam's harbour and demanded its immediate surrender. The following year New Amsterdam was incorporated into the English colonies and renamed New York, after the then Duke of York who later became King James II of England. The English continued to increase their settlements on the east coast, driving out other European contenders, except for the French in the north and the Spanish in the far south.

As farming became increasingly profitable across the English colonies of North America, so grew the demand for land, leading to several important developments. Firstly, the demand for labour, especially in the tobacco and rice plantations of the Carolinas, resulted in the wholesale importation of African slaves. These were supplied by the Royal African Company, founded by the English royal family and a group of London merchants in 1660, with the exclusive right to supply slaves to the English colonies. Secondly, the previously chartered companies and colonial governments claimed huge tracts of land to the west, and since grants of land frequently over-lapped, leading to the same piece of land being claimed by different parties, there were many disputes. Thirdly, land-hungry colonists on their own initiative pushed westwards from the original coastal settlements, occupying lands previ-ously enjoyed by Native Americans. In 1681, to add to this evolving picture, William Penn founded the colony of Penn-sylvania, the idea of which was to encourage mass Quaker emigration from England.

Between 1650 and 1780, as a result of the slave trade, the population of those of African descent grew rapidly in the colonies and became an important part of the so-called trade triangle, in which manufactured goods were taken from England to Africa, in exchange for slaves who were taken to America, in exchange for tobacco, indigo and other goods, which were taken to England. This provided income for British port cities, especially Bristol and Liverpool. In the Caribbean, the African portion of the total population grew to 80 per cent and in the North American colonies, espe-cially in the plantation south, the African population reached 40 per cent.

The English Crown continued to grant royal charters to various companies, aiming to increase the English hold on large regions of North America. In 1670, King Charles II granted a royal charter to the Hudson Bay Company, based on the drainage basin of Hudson's Bay, an area of some 1.5 million square miles in a region then known as Rupert's Land. This company was focused on the exploitation of the fur trade and

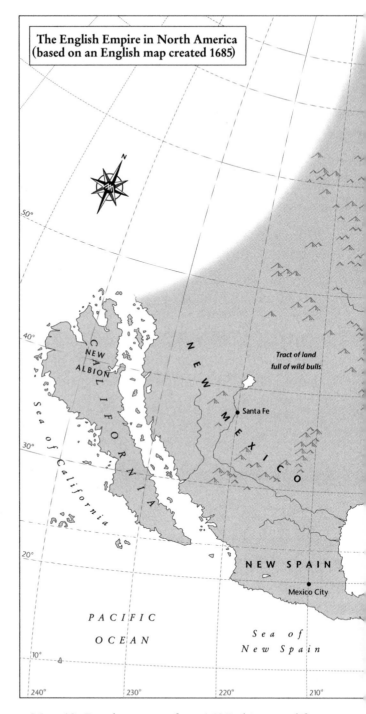

The English Empire in North America
(based on an English map created 1685)

N

50°

40°

NEW
ALBION

CALIFORNIA

Sea of California

30°

20°

NEW MEXICO

Tract of land
full of wild bulls

Santa Fe

NEW SPAIN

Mexico City

PACIFIC
OCEAN

Sea of
New Spain

10°

240°   230°   220°   210°

Map 40. Based on a map from 1685, this exemplifies maps that land grants were based upon. The West was uncharted and California was considered an island.

constructed a chain of forts and trading posts to achieve this end.

In 1695, the Scots decided the time had come for the establishment of Scottish colonies in the New World: 'Och aye, laddie, if the Sassenachs can do it, so can we' was the view held by many canny lowlanders. Consequently, the Company of Scotland was formed by an Act of the Scottish Parliament, guaranteeing a monopoly of Scottish trade to and from India, Africa and the Americas. After the sum of 400,000 Scots pounds had been raised by popular subscription, a plan was developed called the Darien Scheme, which eventually collected somewhere between 20–25 per cent of all monies circulating in Scotland. The idea was to create a trading colony at a strategic point on the isthmus of Panama. Here New Edinburgh would be built; of course, it was already closely surrounded by successful Spanish settlements, and it also threatened existing English interests in the area.

The first expedition of five Scottish ships set out in 1698, and on arrival at the Bay of Darién the settlers set about creating Caledonia, constructing Fort St Andrew. However, the climate was so extreme that farming and hunting proved difficult, and the settlers found it hard, if not impossible, to adapt to their new home. Disease began to take a savage toll and, after just eight months, the survivors abandoned the settlement. The expedition that had been sent to resupply Fort St Andrew found only ruins and it was too late to send news and stop a second expedition from setting out, which arrived in Caledonian Bay in November 1699. The Spanish faced them in their fort constructed nearby and some rebuilding was undertaken at Fort St Andrew in expectation of a Spanish attack. At the same time the mortality rate among the Scottish settlers was horrendous, as it was also among the more weathered Spanish facing them in their fort constructed nearby. After some negotiations, the Scots were allowed to leave New Edinburgh with their guns and equipment, and the colony was finally abandoned for the last time. Out of a total of some 2,500 settlers, only around 400 survived and even fewer ever saw Scotland again.

The reaction to this disaster back in Scotland was profound; almost every family had invested some, or all, of their savings in this scheme and it was a motivating factor in the 1707 Act of Union with England. As Scotland teetered on the edge of bankruptcy, the Scottish elite now considered being part of an existing successful major power the best way out of the consequences of the Darien fiasco. The Union of England and Scotland thus created the Kingdom of Great Britain and the English Empire became the British Empire.

*   *   *

In 1688, a new dynasty came to the English throne in the glory of the 'Bloodless Revolution': James II of England and VII of Scotland was ousted and the Dutch William III, with his wife Mary II, ascended the English throne. This meant that England and the Netherlands became allies and together they entered the Nine Years' War (also known as King William's War in the Americas), a global war that was fought out in Europe and across its colonial possessions.

Next followed the War of the Spanish Succession (also known as Queen Anne's War, just to add confusion), which, once again, was fought out over European and North American possessions. This chaotic bloodbath lasted for 13 years and was concluded by the Treaty of Utrecht in 1713; once again, out of the turmoil, the British Empire regained and gained a number of territories, including Nova Scotia, Newfoundland and other important regions. During the middle decades of the 18th century, localised wars continued on the peripheries of the European colonial possessions, but by 1756 this was largely an Anglo-French struggle, which became the Seven Years' War and concluded with the Treaty of Paris in 1763, confirming Britain's hegemony in North America. What many academics call the First British Empire, which focused on North America and the Caribbean but had growing interests in Africa, Asia and particularly India, was now manifest.

* * *

Many academics have written various articles and studies on the theories of empires – the reasons for their existence, growth and decline. Halford Mackinder's 'Heartland Theory' (1904), for instance, sometimes known as the 'World Island', which demonstrated that the state that possessed the core area of Eurasia could eventually control the world. To put it simply, he stated:

> *Who controls Eastern Europe commands the heartland;*
> *Who rules the heartland commands the world island;*
> *Who rules the world island rules the world.*

The problem with empires is that, despite all of these theories, none of them has ever achieved the projections suggested by the theorists. When examining empires, I noted how they had all come into life in defence of their core. Russia, for example, through protecting its core by conquest and exploitation had spread outwards by 1860 from Poland in the west, to the frontier of British Canada in North America on its eastern border. Russia is a case in point; although it straddles Mackinder's 'Heartland Area', it has singularly failed to become the world's dominant power. Its core territory was attacked twice from the west: first by Napoleon in 1812 and by Hitler in 1941 – both were thrown out on their ears. It fought a brief war on its eastern periphery in 1905 with imperial newcomer Japan and lost – much to the world's surprise. Russia's greatest problems are its vast size and its geography. Its fleets are trapped by ice for a considerable time each year, and its armies are locked down by mountains and millions of square miles of sparsely populated terrain, with little to support further expansion. It was beyond the core area's ability to become the dominating global power.

*   *   *

Back in the British Empire, after 1763 relations between Great Britain and its 13 colonies in North America became increasingly strained, due to the taxation imposed on them by the government in London. The colonies, of course, needed protection, which was provided by the British Government. Our fine lads, dressed in red, formed a shield for the colonies, keeping the vengeful Indians at bay and providing an orderly presence on the streets of settlements across North America. The well-disciplined army and navy could always be counted on. However, this tax was all too much for the colony populations who began to wail and whimper about the rights of Englishmen. Many of the colonials still considered themselves to be Englishmen, or perhaps British, and therefore argued that they deserved representation in the London Parliament, stating, 'No taxation without

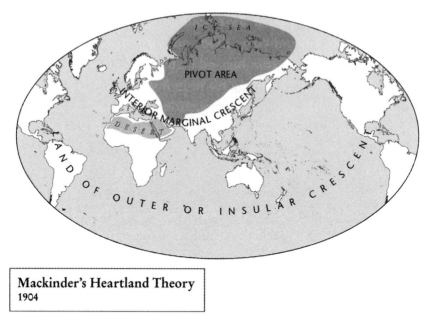

Mackinder's Heartland Theory
1904

Map 41. Halford Mackinder produced his 'Pivot of History' theory in 1904, which extended the scope of geopolitical analysis to encompass the entire globe.

representation.' The British Government, however, responded in a high-handed way by sending troops to support British direct rule, which, in turn, led to the outbreak of war in 1775. For the colonies, the move towards independence had already begun. Despite the fact that many so-called empire loyalists saw themselves as British and that to break with Britain could be seen as treasonous, the bulk of the population had been undergoing a cultural change and were beginning to consider themselves as Americans living in American colonies.

By 1776 the colonies, increasingly feeling independent of Britain's rule, became 'States', and together – United – they declared their independence in 1776. In the ensuing struggle,

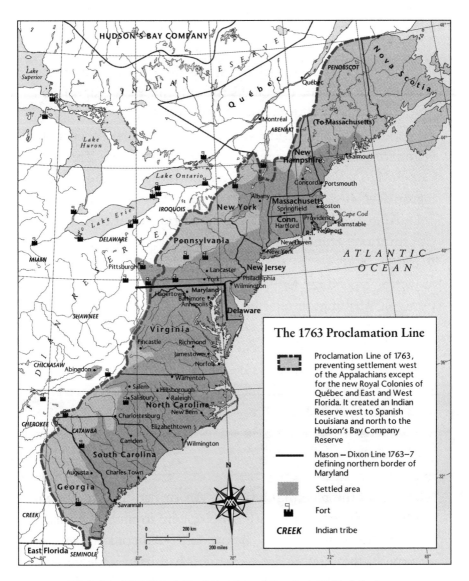

Map 42. *The Royal Proclamation of October 1763 forbade any settlement west of a line drawn along the Appalachian Mountains, which became an Indian reserve.*

both sides were finely balanced, but the entry of France into the war in 1778 favoured an American victory, especially after the siege of Yorktown in 1781, when the colonial Americans with French assistance secured the surrender of a significant portion of the British Army. After this disaster, or victory, depending on which side you took, both parties began to negotiate peace terms, which were agreed by the Treaty of Paris in 1783, and a new nation was born, with George Washington as its first president.

During discussions about the formation of the American Constitution, some new American citizens urged George Washington to accept the title of King of the United States. However, since they had just kicked out one king (also called George), Washington rejected the idea, feeling that the idea of kingship would leave a bad taste in American mouths. Still, as many as 100,000 loyalists decided to leave the newly independent States and make their homes elsewhere within the empire. So ended what some academics call the first (or, as I am calling it, the fourth) British Empire.

*   *   *

Now began the reconfiguration of the British Empire known to some as the second British Empire, and to me as the fifth.* In the 50 years or so before the American War of

---

* I was born at the tail end of the fifth British Empire. That was the British Empire, in its most recent and successful version, bearing in mind how imperial success is judged according to the modern concept of elective self-government and the rights of 'national' groups. The empire into which I had just been born in 1946 was probably at its widest extent, but 20 years later it was almost all gone.

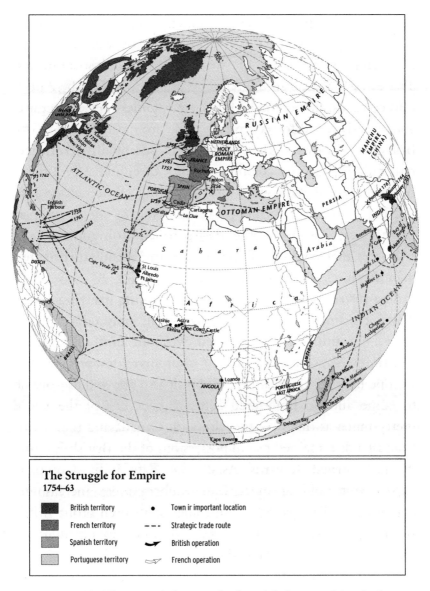

Map 43. This struggle became the first global war and involved most European powers as they attempted to extend their empires around the world.

Independence, Britain had undergone major changes in its economy. The industrial revolution was now well under way and Britain had become the world's leading industrial power. There was a feeling in the country, after the loss of the American colonies, that the need for those colonies was less important than the need to secure free trade around the world. The British now viewed other countries as potential marketplaces rather than potential colonies, though of course it was never that simple – Britain would go on taking one territory after another in a pretty much opportunistic kind of a way, and part of this was driven by the desire to control minerals and resources that could be shipped back to Britain, turned into manufactured goods and re-exported.

The instrument of Britain's power was its growing Royal Navy. During the Napoleonic Wars this only continued to grow, becoming a finely tuned weapon. Not only did it constrain Napoleon's relatively short-lived empire within Europe between 1789 and 1815, but it also allowed the British to come and go unfettered across the oceans of the world pretty much as they saw fit. Although Canada and the Caribbean islands remained useful possessions of the British Empire, its axis turned towards Asia. The East India Company expanded its holdings in the Indian subcontinent, increasingly exploiting India's products and trade for its own profit, while Britain's interests in Southeast Asia continued to expand territorially and grew in financial importance.

After the loss of the American colonies, where previously Britain could banish its miscreants into servitude, Australia became the focus of Britain's attention. It was suggested by James Mario Matra, an American-born empire loyalist of Corsican heritage who had sailed with James Cook to Botany

Bay in 1770, that Australia could be considered as a colony, to be made up of empire loyalists. Instead, the British Government took his idea and adapted it to the notion of transporting its petty criminals there, creating a new colony on the far side of the world. These poor souls were mostly people from the underclass who had been caught holding a slice of apple pie they'd probably forgotten to pay for. James Mario Matra, meanwhile, is remembered in the Sydney suburb of Matraville.

* * *

Surveying Britain's expanding possessions began at home, though the idea of a scientific survey was a French one. In 1783, César-François Cassini de Thury – or Cassini III, of the famed Cassini family, the leading cartographers of their day (see page 169) – contacted the British Government suggesting a cooperative venture. Good God, gasped the Government, what are the Frenchies up to now? Unable to come up with a reply themselves, they passed the problem on to the Royal Society, the British equivalent of the French Royal Academy. Cassini suggested a trigonometric survey to define the exact latitude and longitude of the observatories at Greenwich and Paris, relative to each other. This would then serve to develop a map tying each country together, and country by country a scientifically based map of the world would be created.

Cassini's proposal obviously highlighted the fact that British mapping to date was inferior, or at the very least neglected. After the Royal Society considered the problem, the person they turned to was William Roy. Now, Roy had been part of a detailed survey of the Scottish Highlands, which had been headed by Lieutenant-Colonel David Watson immediately after the Jacobite Rising of 1745–6. The Highlands were a bit of an unknown land to the government forces tasked with rounding up the rebels – the details of the local geography were a little sketchy and the locals spoke Gaelic; as a result, the army frequently arrested – or worse, killed – the wrong people.

Roy went on to enjoy a successful career in the Royal Engineers, achieving the rank of Major-General, and he would shoulder most of the responsibility of the British share of the survey. On previous occasions, he had tried to interest the Government in a general survey of Britain, but it raised no enthusiasm among the politicians. It took the French to do that. Sir Joseph Banks (of Cook's Pacific exploration, see page 147) proposed that Roy undertake the work, and Roy embraced the project. His technical and leadership skills set a very high standard for all that would follow. He submitted an underestimated budget, to say the least, in order to procure the commission; it was difficult to get any progressive investment from the British Government, then as now, but it did at last commit to this long-term, but in the end worthwhile, undertaking.

The Anglo-French survey of 1784–80 began on the British side with the measurement of a baseline, which was fixed on level-ish ground across Hounslow Heath, beginning just under the northern runway of what is now Heathrow Airport and ending at Hampton Poor House, Hampton. During the period when the baseline measurement was being finalised, Roy ordered a new theodolite (a survey instrument for measuring horizontal and vertical angles) from instrument-maker Jesse Ramsden, a procrastinating perfectionist. The letters exchanged between Roy and Ramsden became so harsh that

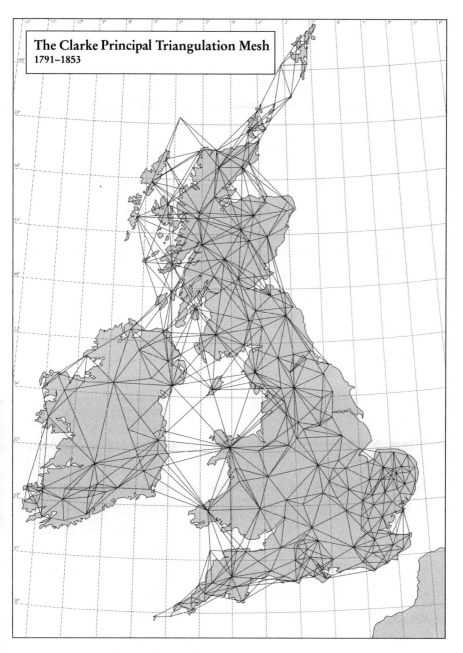

The Clarke Principal Triangulation Mesh
1791–1853

Map 44. The principal triangulation of the British Isles was the first precision trigonometric survey carried out between 1791 and 1853.

they had to be deleted from the Royal Society's final report. Despite the bad language, the theodolite was delivered in 1787 and deployed to immediately rewarding effect.

The survey was completed in 1790 and Major-General William Roy died a little after its completion, on 1 July. In the following year, George III initiated a national survey as a military project. By 1789 the French Revolution had broken out and the Cassinis were badly out of favour with the revolutionaries; war loomed and the United Kingdom needed better maps. In 1801 the first one-inch-to-one-mile, 1:63,360-scale sheet map was published, covering the county of Kent, through which the Anglo-French survey had run. Essex maps followed, setting the style for the Ordnance Survey. The triangulation of the United Kingdom was completed by 1841. In 1858 it was improved by the work of Alexander Ross Clarke, a gifted mathematician, author and officer of the Royal Engineers.

*　　*　　*

Meanwhile across the world, in India, the East India Company set up a mapping department in 1767, which proved very useful for checking up on the exact extent of the company's possessions in India. (I am sure the maps produced looked good illustrating the company's balance sheet.) The first scientific survey of all of India was produced between 1793 to 1796, while the Great Trigonometrical Survey was begun in 1802 and would take 50 years to complete. Lieutenant-Colonel William Lambton began this vast undertaking – incidentally, he would use the Great Theodolite produced by the procrastinating Jesse Ramsden.

Lambton was the prime motivator behind the great trigonometrical survey of India. He was a veteran of the American

Revolution and had moved to India after the American wars to serve under Arthur Wellesley, the future Duke of Wellington. While preparing himself for his new assignment, he studied all the available maps of India and soon concluded that a survey of greater accuracy was required. After being given the necessary permission, Lambton's first action was to order Jesse Ramsden's amazing theodolite. This was duly dispatched from England but the ship carrying the theodolite was captured by the French and taken to Mauritius. Fortunately, it being a gentlemanly era, the French recognised its scientific importance and sent the theodolite on to India.

On 10 April 1802, Lambton began his survey by creating a baseline near the city of Madras, now Chennai, and from there he ran a line of triangulation across southern India, along the line of the 13-degree parallel. His suspicions about the accuracy of previous maps were confirmed and he calculated the width of the Indian peninsula along the 13-degree parallel as 575 kilometres (357 miles) – 70 kilometres (43.5 miles) shorter than earlier maps indicated. This first element of the survey was completed in 1806. The East India Company considered that the project would take about five years to complete, but in the end it took 69 years. Even then, the triangulation did not cover the entirety of India but instead a 'grid iron' of triangulation chains running north–south and east–west. Lambton died at his post in 1823 at the age of 70, working as ever towards the completion of the great survey. His assistant, Colonel George Everest, took his place and continued the project for another 20 years.

By 1834 the survey had reached the foothills of the Himalayas, and over the next nine years Everest extended it well into the mountain range. On the border of Nepal and Tibet there was a majestic mountain, simply known as Peak XV. Everest retired in 1843 and it is not known whether he even saw this great mountain that would eventually bear his name. Due to its location, Peak XV was forbidden territory to the Indian survey. However, in 1849, the surveyors climbed lesser summits and took theodolite readings of the majestic peak from six different directions. Given the technical and physical problems of the time, it is not surprising that none of the measurements agreed. A Bengali clerk checked all the results and by averaging out the measurements came up with a height of exactly 29,000 feet (8,839.2 metres). Apparently he rushed into the office of Everest's successor, General Andrew Scott Waugh, and presented him with the information. The clerk, whose name unfortunately has been lost in history, blurted out, 'Sir, I have discovered the highest mountain in the world!' Waugh considered this for a moment and, over the next couple of years, carried out a number of checks, which included asking Radhanath Sikdar, an Indian mathematician, to check the survey's findings and come up with a final calculation as to the height of Peak XV. Eventually Sikdar's findings came to 29,000 feet (8,839.2 metres), the same as the Bengali clerk's. Waugh considered this measurement to sound like a rounded estimate and therefore added 2 feet (0.6 metre) to the total. The survey had previously used local names for the physical features covered in the survey; however, this particular mountain had been identified several times using different names. Therefore, Waugh argued that it was impossible to choose a name in constant use. Instead, he chose to name the mountain after his predecessor, Everest.*

---

*I should tell you here that George preferred his family name to be pro-

Both Waugh and the survey survived the end of the East India Company when it ceased to exist after the Indian Rebellion of 1857 and the British Government took direct rule over India. The government also took over the great trigonometrical survey of India, which was eventually completed in 1871. It was the most extensive cartographic survey that had ever been attempted, and one of the most difficult, carried out under the most testing conditions. Even when it was concluded, the larger task of mapping all of India was still to be completed. However, a standard of excellence had now been put in place.

$$* \quad * \quad *$$

Between the end of the Napoleonic Wars and the beginning of the First World War, somewhere around 26 million square kilometres (10 million square miles) of territory had been added to the British Empire; these areas contained about 400 million people. As well as direct control over these territories, Britain's prominent position in world trade meant that it effectively controlled the economies of many other countries; for instance, Argentina, Siam (now Thailand) and large regions of China. Britain's global dominance was also still backed by the might of the Royal Navy. This had evolved from the wooden warships of Nelson's day, to the steam-driven iron clads of the 1860s and the newly developed armoured warships of the early 1900s, led by the scientific wonder of the day: the dreadnought battleship, the first of which was launched in 1906. This blatant

nounced 'Eve-rest', not 'Ev-er-est' as we pronounce it nowadays. So next time you're in a pub quiz, make sure you get the name of the highest mountain correct.

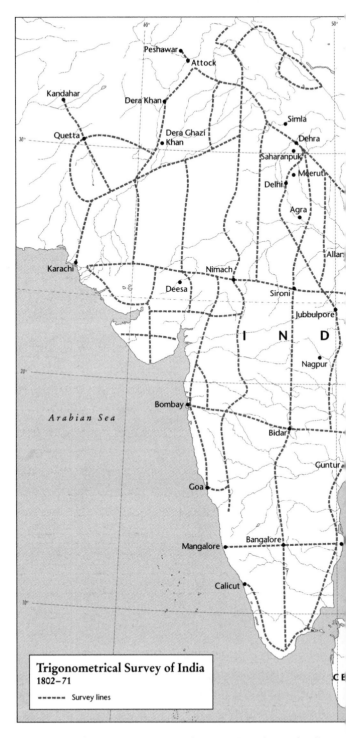

Map 45. *The survey was started in 1802 and was finally completed in 1871, giving the British administration an insight into their territories within the Indian subcontinent.*

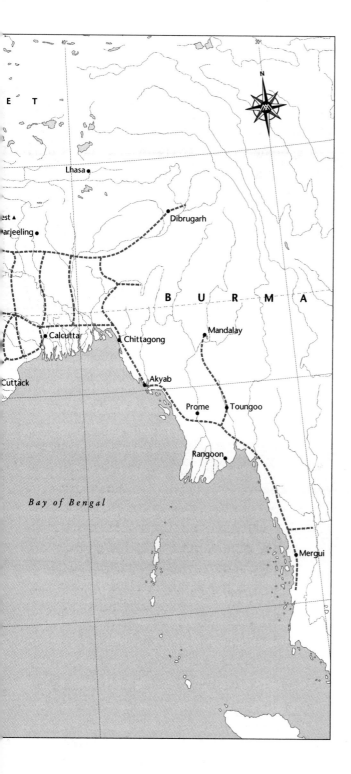

example of power projection protected a gigantic merchant fleet, which dominated the sea lanes that connected an empire 'upon which the sun never set'. This empire had seen off its Spanish, Dutch and French imperial rivals, and successfully contained the Russian Empire.

14

GONE WEST

I suppose I was about 10 years old when I decided I wanted to go to the United States. Of course, I had a dream of myself in later life heading out on a steamer from Liverpool bound for New York. My dream allowed me to travel in something better than steerage class; the crossing of the Atlantic Ocean wouldn't be too storm-tossed and I would arrive in New York Harbour, gliding past the Statue of Liberty, gripping the ship's rails and straining for the first view of Manhattan's towers. In fact, I was in my early thirties and the journey didn't take days but almost exactly seven hours. I flew from Manchester Airport to Boston, Massachusetts, with the idea of comparing New England to the old England I knew so well.

I travelled with 30 years of preconceived ideas about the country I was visiting. I had read scores of books; I knew about pre-Columbian migration with Clovis spearheads, the early Spanish explorers, the first English settlements along the east coast (including settlements not far from Boston), the early Pilgrims meeting with the Native Americans, the struggles with Dutch settlers who were eventually overwhelmed and incorporated into the English colonies, and much more – or so I thought. I sat on the plane clutching my copy of *Voyagers to the West* by Bernard Bailyn, a well-thumbed, dog-eared copy; I somehow thought that would prepare me for the New World. The other guide that helped to form my ideas was the weekly talk by Alistair Cooke on BBC Radio called *Letter from America*. Of course, all this prepared me for absolutely nothing, and my mental maps of the US were of no avail.

Leaving Boston, I chose to travel on to New York by train; it was about a four-hour journey and I hoped it would deliver a view of a slice of New England. We glided out of the station and into the countryside of Massachusetts. According to the maps I had studied at home, we should be going through farmland, quite a lot of it given over to dairy, but all I could see in all directions were thousands and thousands of acres of forest. Occasionally, as the train rolled along, I could see the remains of stone walls snaking off into the trees and I thought, Why would anyone bother to build field walls through the woods? I later found out from Hammond, the cartographic publishers based in New Jersey, that somehow their maps had never been updated since the 1920s and what happened was this: the hard-headed farmers of New England had been moving west since the 1850s, and by the early decades of the 20th century, they had gradually abandoned the stony ground of New England for better land in Ohio, Iowa and other parts west. So, what had been open farmland had slowly reverted to forest. Looking at the towns and landscapes of New England, I could see straight away that the settlers from old England found themselves in a very difficult terrain to cultivate, even without the July temperatures – it was hot, at least by my standards. We passed through New Haven, Fairfield and Greenwich, eventually into downtown Manhattan and finally into Penn Station.

Over the next two years following this first foray, America blossomed into being the largest market for my company's maps. Almost half of everything we produced from then on covered American topics. The flight between Manchester and New York became a regular run, as did journeys to Washington to visit the Smithsonian Institution and the National Geographical Society, where we planned an *Atlas of American History*, which eventually would be published by another company.

A couple of years later, after completing a commission for a client in Washington, DC covering colonial America, I decided on a few days off before taking on the next job in New York. My time would be spent on a slow drive north, taking in the lands I had mapped, or a part of it, over the last few months. Heading north on country and local roads, I passed names that were familiar (Westminster, Hampstead), names that spoke of hope (Libertytown) and of founders (Germantown), before drawing close to the Maryland and Pennsylvania border. Crossing this line – the Mason–Dixon Line – would mean leaving 'Dixie' for the north. This was one of those rare moments when mapmakers not only defined a line on the map, but also made a true cultural mark in history.

Taking a long look at the well-settled landscape around me, it was not so easy to imagine the land in the 1760s – an extraordinary period just after the end of the Seven Years' War, which was fought in North America as the French and Indian War of 1756–63. In the years of peace following 1763, transatlantic migration from the British Isles and, to a lesser extent, northwestern Europe, reached levels beyond anything previously seen in the British North American colonies. Arrivals in the Pennsylvania, Maryland and Delaware regions were more than half of all the arrivals in the Americas. As the population began to rise rapidly and new settlements formed, the inaccurate maps based on earlier land grants issued by Britain began to cause land disputes in the colonies.

In 1632, King Charles I of England had awarded the first Lord Baltimore, George Calvert, letters patent 'for all the land – hitherto unsettled – from the Potomac River north to the 40th parallel and then running westward from the Atlantic Ocean to a meridian through the first mountain of the Potomac'. Wherever that might transpire to be! Alas, the first Lord Baltimore died a couple of weeks before the paperwork was issued, and so his son, Cecil, the second Lord Baltimore, took up the challenge to define the location of the correct boundary.

Half a century passed by, the British Isles passed through a vicious civil war and Charles I lost his head before an approving crowd of Londoners on a cold January day in 1649. It was not until May 1660 that the monarchy was restored when Charles II was invited to ascend the throne and, in 1681, Charles granted a charter to William Penn, giving him title to a territory north of Maryland that was bounded to the south along the same 40th parallel as Maryland's northern border.

The problems arose out of the inaccuracy of the maps upon which both of these charters were drawn up. When it came to actually pegging out the claims later, surveyors discovered that the 40th parallel was too far south. Lord Baltimore compounded this misinterpretation of geography when he commissioned his own map in 1635, by accepting much of the 1608 map. This did not help the Calvert claim, locating his own northern border below the upper end of Chesapeake Bay. To add to this dispute, in 1682 the Penn family had acquired additional lands on the western shore of Delaware Bay. The charter issued by Charles I concluded that, since these lands were already settled by Dutch and later Swedish Christian folk, the Delaware counties were excluded from the Calvert patent.

Decades of disputes, counter-claims and litigation began when the third Lord Baltimore, Charles Calvert, led a party of officials – all well armed and taking care to measure the latitude as they proceeded – up the Delaware River to instruct the citizens of Chester, Marcus Hook and Newcastle to pay taxes to him and not to the Penns. Pennsylvania officials also made taxation claims of their own and later confounded a certain Thomas Cresap. Born in Yorkshire but settled on the banks of the Susquehanna River, Cresap was a successful trader who managed to have a war named after him after executing his duty as agent for Lord Baltimore in what was considered Penn territory. In 1730, 'Cresap's War' was triggered after skirmishes and property theft took place in the disputed areas. Maryland called out its militia in 1736; Pennsylvania reciprocated in 1737. Frontier ruffians from both sides, or those with no particular allegiance, used the opportunity to relieve remote farmsteads of their goods and chattels.

In 1738, King George II forced both parties in his unruly colonies into a ceasefire. This edgy situation was partially resolved in 1750 by a judgement made by England's senior judge, the Lord Chancellor, which was accepted by the Calverts and the Penns in 1760. The new border would run west 15 miles south of the city of Philadelphia, and the north–south Delaware border was also agreed. However, it was one thing to draw lines in London, but another entirely to fix these lines on ground often covered by dense woodland, wide rivers and swamp. Various local surveyors proved unsatisfactory to both the Calvert and Penn camps, and so eventually they consulted the Astronomer Royal in London as the source of neutral and accurate arbitration. The president of the Royal Society suggested two men for the task: Charles Mason from Gloucestershire, southwest England, and Jeremiah Dixon, born in County Durham in the northwest of England.

It was this pair of experienced men who arrived in Philadelphia in November 1763, bringing with them the latest instruments and – perhaps more importantly for the Calverts and Penns – an unbiased approach. Before the survey could begin, the Calvert and Penn parties had concluded that the border should, following the words of the charter, be 'under the 40th parallel', which meant 15 miles south of the southernmost limits of the city of Philadelphia. The surveyors first fixed this point at 39 degrees 56 minutes 29.1 seconds of latitude (subsequent observations show this to be off by just 2.5 minutes). They then moved west, along this line of latitude, to avoid New Jersey territories, their instruments carried by a wagon upon a feather bed. They arrived at the farm belonging to the Harland family, who originated from northeast England, as did Dixon himself. On the Harland farm, they took their readings, setting up a marker stone known to the Harland family and other locals as the 'stargazers' stone' – it is still there to this day. The surveyors moved on to a point 26 kilometres south, which landed them almost in the front garden of a Mr Alexander Bryant. Mason and Dixon proposed that it was at this spot that the survey should begin, and agents of the Penns and Calverts agreed.

Between June and September 1764, they worked on the north–south line in the Delmarva Peninsula between what would become the colonies, and later still the states, of

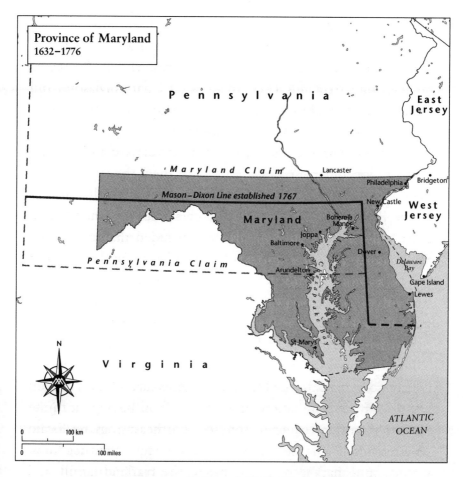

*Map 46. Between 1763 and 1767 Charles Mason and
Jeremiah Dixon surveyed a line in Colonial America that not only
solved territorial issues between Maryland, Pennsylvania and
Delaware, but also created a cultural dividing line.*

Delaware and Maryland. With the survey partly led by a group
of axe men, they hacked their way 135 kilometres south,
leaving a waist-high limestone marker stone every mile of the
way. Mason and Dixon returned to the Bryant house in the
early spring of 1765 and here they placed their first marker

stone of the east–west line, which they described as the 'post marked West'. Their work, as ever, was meticulous; the axe men cleared a path around 10 metres wide to allow for clear observation, frequently checking and retracing their steps and measurements, and they also used a Gunter's chain made up of 66 links to measure over level ground. They took constant astronomical observations every 10 minutes of degree (about every 17.5 kilometres).

By October 1765, Mason and Dixon had reached the summit of the Blue Ridge Mountains and they then returned to the coast for the winter. In spring 1766, they returned to the line, placing limestone markers in place of the timber markers constructed the previous year and extending the line westwards. They reached a point 5 degrees of longitude west of the starting point, as indicated by the Pennsylvania charter, thereby completing their initial task.

During the winter break of 1766–7, the surveyors had the idea of going back to the Delmarva Peninsula and carrying out a geodetic determination (a method of achieving accurate measurements by using astronomical observations to locate precise reference points on the earth's surface) of one degree of latitude. This had been done in other locations (Peru, Finland), but so far not in the North American colonies. With the backing of the Calverts and Penns, who thought this would help in refining local maps and claims, they went ahead.

While this was in hand, officials of the colonies had come to an agreement with the Six Nations Indian tribes to continue the survey westwards. Mason and Dixon were urged to treat the tribes well for their own protection, and this included ample servings of liquor to be handed out 'not more than three times a day'. During the summer of 1767, as the extended survey pushed westwards, the party's Indian guides began to disappear and the whole group began to feel more and more exposed. As they reached Dunkard Creek, the axe men and others refused to go on. Here Mason and Dixon took their final measurements, which revealed they were 233 miles, 3 chains and 38 links from the 'post marked West', a point now on the Pennsylvania and West Virginia border.

In September 1768, Mason and Dixon returned to England. Jeremiah Dixon died in 1779. Charles Mason continued to work, but after a physical and financial decline, and having returned to Philadelphia in 1780 without explanation, he died there in 1786 and was buried in an unmarked grave. Mason's and Dixon's names live on as a line on the map that represents so much – a cultural divide between north and south, freeman and slave, rebellion and union: the Mason–Dixon Line.

15

A WORLD AT WAR

As a mapmaker, it seems warfare has paid my mortgage, put food on the table and educated my children. I have spent more of my creative life working on this subject than any other. One of my earliest undertakings was *The Atlas of the Crusades* with the late, wonderful Professor Jonathan Riley-Smith, whom I met in the mid-1980s when he was at the University of London.

The origins of the Crusades began in March 1095 when an embassy from the Christian Byzantine Empire asked Pope Urban II for help against the Seljuk Turks who had overrun its eastern provinces. Urban II had already decided that he would appeal to the knights of Western Europe to come to the aid of these Christian communities; he had developed the idea of summoning the knights to take part in a war that was also simultaneously a pilgrimage to Jerusalem: this turned out to be the First Crusade.

The pope had launched a religious movement that only subsided in the late 18th century. In many respects, the Crusades themselves had faded in the European memory by then, but in the Middle East they had remained in the popular memory. For instance, one of the links that lingered into the

Map 47. *The Crusades, sanctioned by the Catholic Church, were Christian Europe's first military intervention outside the bounds of the continent.*

modern period was the connection forged by the indigenous Maronite Christians of the Near East and the Church of Rome to defend the holy places of Christendom; many Maronites regard the period of the Crusades as a 'Golden Age'. While in the 1950s, the president of Egypt, Gamal Abdel Nasser, compared the French and British imperialist attempt to seize the recently nationalised Suez Canal to Saladin's struggle with the Crusaders.

From the 14th to 15th centuries, at the height of the crusading period, hundreds of thousands of men and women, and on one occasion children, took to the road heading east, armed with the theology of sacred violence in which the perpetrator must have the right intention and just cause in response to a previous injury, which together legitimised their actions. Many of those on crusade were armed with their belief and little else, dying by the thousands en route. All the followers of the cross were granted various favours to go on crusade, from tax exemptions to pardons for past crimes and, of course, a place in heaven. This outpouring created the Christian states of the Near East. Some consider this Europe's first attempt at expansion since the fall of Rome.

Against all odds, the First Crusade succeeded in capturing Jerusalem in 1099 in the usual welter of blood, and the early Christian strongholds formed into a number of small states, which lasted until 1302 – the year in which the last two strongholds, Ruad and Gibelet on the coast of the County of Tripoli, fell. The crusading impulse lost its impetus in 1378 when the Catholic Church split in the 'Great Schism' of 1378. Europe was now torn between two rival popes – one in Rome and one in Avignon – and both were happy to mobilise crusades against each other. It wasn't until 1417 that the election of

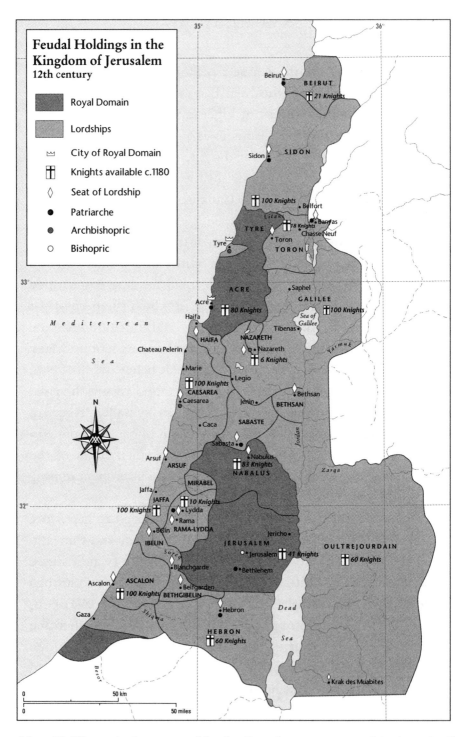

Feudal Holdings in the Kingdom of Jerusalem
12th century

- Royal Domain
- Lordships
- ⌂ City of Royal Domain
- ♰ Knights available c.1180
- ◇ Seat of Lordship
- ● Patriarche
- ◉ Archbishopric
- ○ Bishopric

35°
36°

Beirut
BEIRUT
♰ 21 Knights

Sidon
SIDON

♰ 100 Knights
Belfort
Lilan
Banyas
TYRE ♰ 18 Knights
Chasse Neuf
Toron
Tyre
TORON

33°

ACRE
Saphel
Acre
GALILEE
♰ 80 Knights
♰ 100 Knights
Haifa
Sea of Galilee

Mediterranean
Tibenas
HAIFA
NAZARETH
Chateau Pelerin
Nazareth
Sea
♰ 6 Knights
Marie
Legio
♰ 100 Knights
CAESAREA
Bethsan
Caesarea
Jenin
BETHSAN
Caca
SABASTE
Jordan
Sabasta
Arsuf
Nabulus
ARSUF
♰ 83 Knights
Zarqa
NABALUS
Jaffa
MIRABEL
JAFFA
♰ 10 Knights
32°
100 Knights ♰
Lydda
Rama
Belin **RAMA-LYDDA**
Jericho
IBELIN
Safe
JERUSALEM
OULTREJOURDAIN
Jerusalem ♰ 41 Knights
♰ 60 Knights
Blanchgarde
Bethlehem
Ascalon
ASCALON
Belfgarden
♰ 100 Knights **BETHGIBELIN**
Gaza
Hebron
Dead
Shiqma
HEBRON
Sea
♰ 60 Knights
Besor

N

0 ___ 50 km
0 ___ 50 miles

Krak des Muabites

Map 48. The territories conquered by the Crusaders were structured in the style of medieval Europe, with each feudal holding supplying knights to protect the state.

Pope Martin V reunified all of Europe. By then, however, the growing power of the Muslim Ottoman Empire had expanded into the Balkans, cutting off what was left of the Byzantine Empire – little more than a patch of land around the city of Constantinople. By the time the Spanish reconquests in the West were complete, with the fall of Granada in 1492, the Ottoman conquests in the East had reached Belgrade. Ottoman expansion continued, reaching the gates of Vienna in 1683; at this moment of peril the Christian states of Europe set aside their differences (they usually had plenty to argue about) and formed the Holy League to fend off this invasion, which they partly achieved, driving the Ottomans back to the line of the Danube River, where the border would remain until the 1900s. It seems that religious wars are the meanest and least forgiving of all wars, particularly when there are two interpretations of the same religion, as demonstrated in the struggles between Catholics and Protestants in Christendom and Sunnis and Shias in Islam.

*   *   *

While designing and producing *The Atlas of the Crusades*, another project appeared on the horizon; this was to be an atlas on the American Civil War. Out of the history of the United States, certainly by the late 1850s, two distinct societies had developed, and both had divergent economic objectives. The South, with its low-cost, slave-labour economy, needed free trade to export its major cash crop – cotton. Meanwhile, the North continued to industrialise, absorbing large numbers of immigrants from Europe in the process. It was this economy that needed some kind of trade protectionism to secure its growing industries.

After 1848 and its successful war with Mexico, the United States added huge new territories to the republic. The Southern slave-owners, of course, wanted to extend their activities westwards and demanded the right to take their slaves into the new territories. The Northerners equally demanded that slavery be restricted and that the new territories should be 'free territories'. Various compromises, cobbled together in the 1850s, really began to give way with the election of Abraham Lincoln in 1860. He gave his inaugural address on 4 March 1861, in which he went out of his way to reassure the eight slave states that had not yet seceded and cool the temper of the seven Southern Confederate states that had already done so. He said it was not the federal government's remit to interfere with the ordering of the individual state's societies, including slavery.

On 11 April 1861, Confederate Brigadier-General P.G.T. Beauregard demanded the surrender of the federal Fort Sumpter, located in Charleston Harbour, and the fort refused to comply. On 12 April, the Confederate artillery bombarded the fort, which returned fire, and on 13 April Fort Sumpter surrendered. A previously hesitant North was now united behind Lincoln in his mission to preserve the Union.

The first Battle of Bull Run, as it was known in the North, or the first Battle of Manassas, as it was known in the South, was the first major clash of the civil war and would disillusion those hopeful souls, especially in the South, who believed that they could quickly win this war by decisive action. Its bloody outcome demonstrated that this was going to be a long war of attrition, which the South could not win.

At the beginning of the war, the then Federal or Union Army was just 16,367 strong, including 1,106 officers, most of whom were Southerners. Many of the latter broke their oath to the Union and left for their states of origin to help form the Confederate States Army, which left the then North with a very small core with which to build an army to save the Union. The president began by ordering the states to raise 75,000 men within three months, a call that forced four more states to join the Confederacy. Worrying as that seemed, the real power still lay with the North, as anyone with a pencil and the back of an envelope could work out. The North had a population of 22 million, the South 9 million, of which 3.9 million were slaves. The North, with 120,000 factories, produced all its military and civil needs; the South, with 20,000 factories, produced some items in limited quantity and relied on imports to meet a large part of its military needs. The North would eventually lose 360,000 dead but still have 1,000,000 men under arms at the war's end. Its population and industrial capacity increased. The South lost 260,000 dead – one in four white men of military age – and almost 70 per cent of its wealth was destroyed.

Confederate and Union armies grew rapidly; the tactics deployed had changed little from the time of the Napoleonic wars and the war with Mexico, but (and this is a major but) the old smoothbore musket with its round musket ball had been replaced by rifle muskets, which were carried into battle by the infantry. The rifling or spiral groves that ran down the barrel fired a mini bullet that expanded into the grooves, which imparted a spin, increasing the shot's accuracy by a large margin. This meant that as two parallel lines of infantry faced each other, they could open accurate fire on the enemy at a much greater distance. Instead of firing just once, then

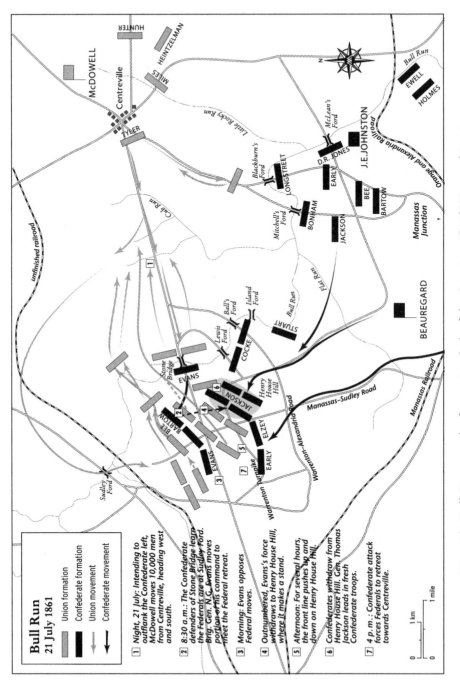

Bull Run
21 July 1861

Union formation

Confederate formation

Union movement

Confederate movement

1 Night, 21 July: Intending to outflank the Confederate left, McDowell moves 10,000 men from Centreville, heading west and south.

2 8:30 a.m.: The Confederate defenders of Stone Bridge learn the Federals are at Sudley Ford. Brig. Gen. N.C. Evans moves portion of his command to meet the Federal retreat.

3 Morning: Evans opposes Federal moves.

4 Outnumbered, Evans's force withdraws to Henry House Hill, where it makes a stand.

5 Afternoon: For several hours, the front line pushes up and down on Henry House Hill.

6 Confederates withdraw from Henry House Hill. Gen. Thomas Jackson leads in fresh Confederate troops.

7 4 p.m.: Confederate attack forces Federals to retreat towards Centreville.

0 ___ 1 km
0 ___ 1 mile

Map 49. Bull Run was the first major clash of the American Civil War, resulting in a fearsome rate of loss among the combatants.

charging with the bayonet, they could fire three or more times, creating horrific casualties. These would only get worse when repeating rifles became available. The Confederate soldiers who took part in Pickett's Charge at the Battle of Gettysburg lost half of their attacking force before they could retaliate; brave as they were, by then it was too late to be effective and the Confederate invasion of the North was lost. Those desperate moments in civil war battles were not unusual. The waves of Union soldiers attacking the stone wall at Fredericksburg the year before lost 80 per cent of their attacking infantry, yet, faced with horrific losses, both sides would not give way. Such was the commitment to the battle's outcome.

After the Battle of Gettysburg on 1–3 July 1863, President Lincoln gave a speech known as the Gettysburg Address that was intended, in many ways in the midst of war, to begin to heal the terrible wounds inflicted on the memory of all Americans.

After the end of the American Civil War and before the outbreak of the First World War, the world was at war almost continuously. During this period, a system of allegiances was established in Europe – newly unified Germany created a dual alliance with the Austro-Hungarian Empire, which was extended to Italy in 1882 and was known as the Triple Alliance, which was later supported by Romania. This central block sat astride the continent. Alarmed by Germany's rapid economic growth, France looked to create an alliance with Imperial Russia, before establishing a neutrality agreement with Italy and finally, in 1904, signing the Entente Cordial peace treaty with Britain. In turn, Britain had a long-standing arrangement to protect the neutrality of Belgium and had

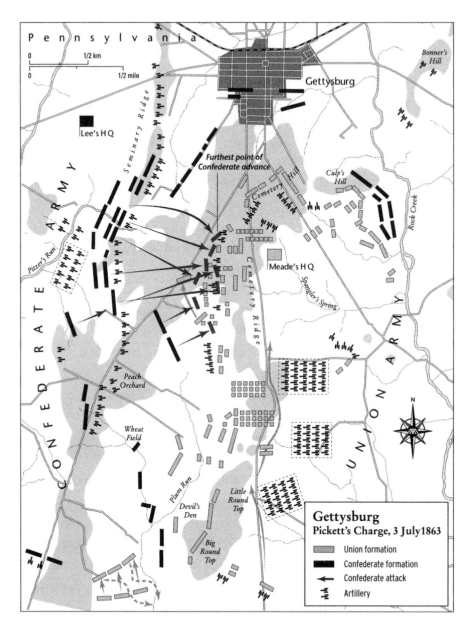

Map 50. *The long approach taken by General Pickett's Division of the Confederate States Army exposed his advancing troops to cannon and, later, musket fire, creating horrendous casualties.*

entered into a conciliation of interest with Russia, and a secret treaty with Italy.

This system of alliances largely worked until after 1897, when Germany aspired to seek 'a place in the sun'. This desire to create an empire was a clear threat to the older, established imperial powers. Germany had also set about creating what it liked to call the High Sea Fleet, which would rival the Royal Navy. In Britain, this was the main driving force in coming to terms with France in 1904 and Russia in 1907. The French, in their turn, ended their disputes with Italy in 1902. Thereafter the Triple Entente with France, Russia and Britain represented the foreign policy of these three countries with respect to the perceived German threat. Germany, in its turn, became concerned by the danger of encirclement, largely created by Britain.

The temperature in Europe was slowly being turned up and revolved around two armed camps – the Triple Alliance and the Triple Entente. On the southern frontiers of the Triple Alliance lay the territory of the Ottoman Empire. By the latter part of the 19th century, the Ottomans were a failing power. By 1878, Serbia, Romania, parts of Bulgaria and much of Greece were now independent of the Ottomans and all the new states were eager to establish themselves and expand if possible, which was especially noticeable in Serbia. The Austro-Hungarian Empire had moved into Bosnia-Herzegovina in 1878 and, with German backing, annexed the territory in 1908. This did not please everybody, especially Serbian nationalists. It also stimulated antagonism between the Austro-Hungarian Empire, itself a multi-ethnic state, and Russia. Russia regarded itself as a kind of protector of the Balkan Slavs and had sponsored the Balkan League, aimed at driving the Ottomans out of Europe. The Balkan Wars of 1912–13 had partially achieved this objective. Serbia doubled in size, which in turn created some alarm in Austria. On 28 June 1914, a Serb living in Bosnia, Gavrilo Princip, a member of Young Bosnia – an organisation seeking to end Austro-Hungarian rule – assassinated the heir to the Austrian throne, Archduke Franz Ferdinand, in Sarajevo.

The Austrian Government issued an ultimatum to Serbia; it seemed like its last chance to deal with the upstart country before it became too powerful. Serbia's reply failed to satisfy the Austro-Hungarians and both countries began to place themselves on a war-footing. Germany stood behind its close ally. Russia, in its turn, could not afford to be seen to leave Serbia unprotected and began to mobilise against the Austro-Hungarians. Germany regarded Russian mobilisation as a direct threat to its security and began to mobilise. The country's grand strategy called for an immediate attack on France, Russia's ally, and the hope was that France would quickly be defeated, before Russia, with its army spread across huge distances, could mobilise to its western border. On 28 July, Austria declared war on Serbia; on 1 August, Germany declared war on Russia and, on 3 August, on France. On 4 August, Britain declared war on Germany in compliance with its alliances with Belgium, France and Russia. The system of alliances, together with a volatile situation in the Balkans, had run out of control and precipitated what would come to be called the Great War.

The war plans of the great powers, such as they were, envisaged a short, decisive affair. The German 'Schlieffen Plan' (a war plan named after Field Marshal Alfred von Schlieffen) was based on a swift advance through Belgium, wheeling south, curving around Paris and knocking the French out of the war quickly, before Russia could mobilise and concentrate its forces. At first, the plan seemed successful as the German armies crossed Belgium and advanced deep into France, but they failed to envelop Paris. A determined counter-attack on the part of the French and British armies along the Marne River succeeded in stopping the offensive and pushed the Germans back slightly, but both sides decided to dig in, extending their flanks until a line of trenches extended from the North Sea to the Swiss border.

Meanwhile, in the east, Russian armies mobilised a little faster than the German timetable had allowed and attacked East Prussia. The Germans defeated this attack at Tannenberg in August, which was done by diverting troops from the Western Front.

The war along the Western Front turned into a long war of attrition, interspersed with major offensives. The defensive firepower of bolt-action rifles, machine guns and quick-fire artillery, together with infestations of barbed wire, made a breakthrough almost impossible. Nevertheless, huge battles that could last many months were fought out: Verdun (February to December 1916); the Somme (July to November 1916); Passchendaele (July to November 1917). The availability of a complex rail network in Western Europe, together with motor transport, also allowed the fast movement of reserves to any threatened point.

The Allies attempted to find ways around this deadlock and attacked the Ottoman Empire, which had entered the war on the side of the Central Powers, namely Germany and Austria-Hungary, in October 1914. Britain initially attacked across Suez and landed a force in Mesopotamia in 1914/early

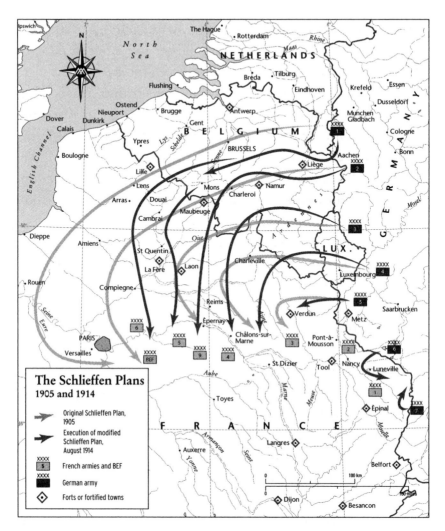

The Schlieffen Plans
1905 and 1914

→ Original Schlieffen Plan, 1905

→ Execution of modified Schlieffen Plan, August 1914

XXXX / 5 French armies and BEF

XXXX German army

◈ Forts or fortified towns

Map 51. The Schlieffen Plan was intended to be a swift-moving operation to knock France out of the First World War. However, reality proved somewhat different.

1915 and in February 1915 launched a naval attack in Gallipoli, which was ultimately unsuccessful. The Allies opened another front in Salonica, northern Greece, in October 1915 in an attempt to support Serbia. Most of these fronts made little progress, with the exception of Palestine, where support for the Allies from the Arab Revolt helped in the capture of Jerusalem and Damascus.

In May 1915, Italy declared war on the side of the Allies and opened a front against the Austrian Empire along the line of the Alps. Numerous offensives were launched by Italy but made little progress. The Central Powers attacked in October 1917, pushing the Italians back to the line of the Piave River, where the Italians made a stand. The Battle of Vittorio Veneto, fought from 23 October to 1 November 1918, saw the Austro-Hungarians comprehensively defeated at last on the Italian front.

Germany, meanwhile, was suffering the effects of an Allied naval blockade from the end of 1916. The German High Seas Fleet launched its only major operation to confront and destroy at least part of the British Grand Fleet in the North Sea. On 31 May 1916, the two fleets met off the coast of Jutland. In the ensuing battle the German fleet was driven back to its ports; the British lost more ships but remained in possession of the sea. In an attempt to counter this, Germany launched a submarine war in the Atlantic Ocean with the intention of cutting Britain's lines of supply, upon which Britain's war effort relied. However, in 1917, unrestricted submarine warfare failed in its objective and the sinking of American ships provoked the United States to declare war on Germany. The war was now being fought out in German colonies in Africa and the Pacific, and on the coast of China, as well as in Europe and the Middle East.

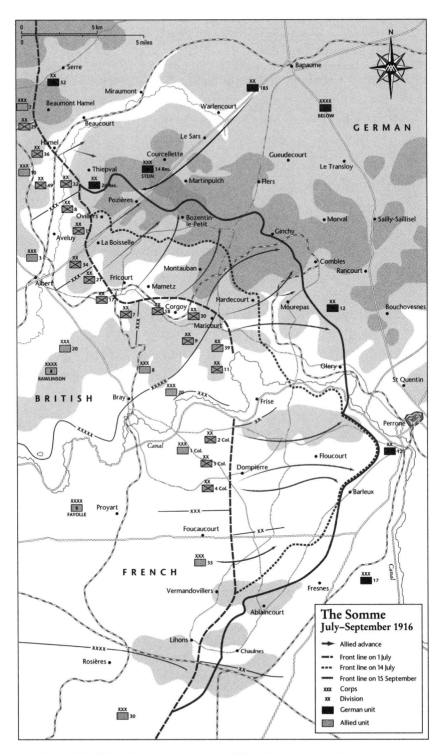

Map 52. The plan for the Somme Offensive was created in immense detail. However, the outcome proved this time that defence would always overwhelm offence.

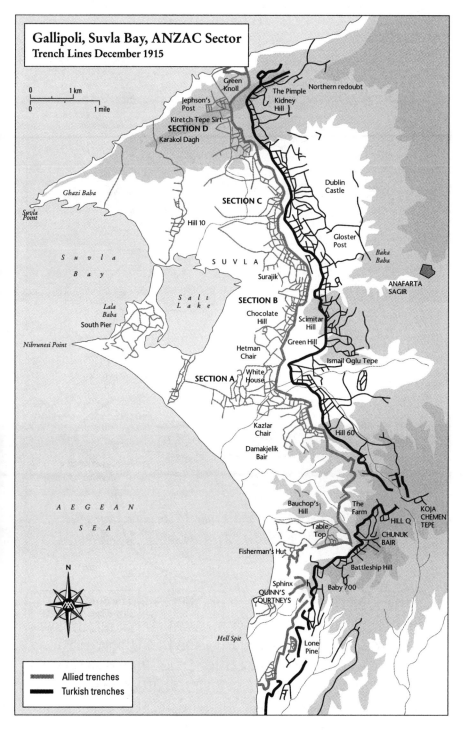

Gallipoli, Suvla Bay, ANZAC Sector
Trench Lines December 1915

0 ——— 1 km
0 ——— 1 mile

Green Knoll
The Pimple
Kidney Hill
Northern redoubt
Jephson's Post
Kiretch Tepe Sirt
SECTION D
Karakol Dagh
Ghazi Baba
Suvla Point
SECTION C
Dublin Castle
Hill 10
Gloster Post
Baka Baba
S u v l a
B a y
S U V L A
Surajik
ANAFARTA SAGIR
S a l t
L a k e
SECTION B
Lala Baba
Chocolate Hill
Scimitar Hill
South Pier
Green Hill
Nibrunesi Point
Hetman Chair
Ismail Oglu Tepe
White House
SECTION A
Kazlar Chair
Hill 60
Damakjelik Bair
A E G E A N
Bauchop's Hill
The Farm
KOJA CHEMEN TEPE
S E A
HILL Q
Table Top
CHUNUK BAIR
Fisherman's Hut
Battleship Hill
N
Sphinx
Baby 700
QUINN'S
COURTNEYS
Hell Spit
Lone Pine

Allied trenches
Turkish trenches

*Map 53. The Allies failed to gain a substantial bridgehead at Gallipoli;
Ottoman forces were able to establish defensive trench lines, leading to an
eventual stalemate.*

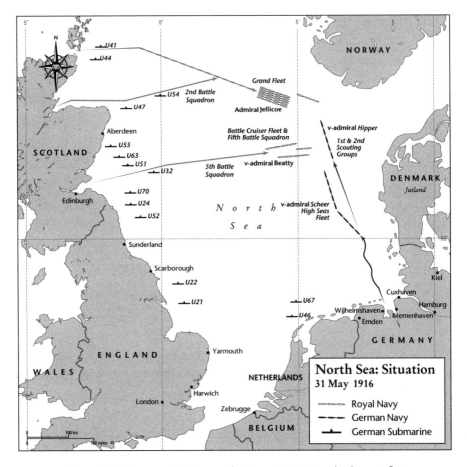

Map 54. Between 31 May and 1 June 1916, in the largest fleet action of the First World War, a 151-strong British fleet defeated a 99-strong German fleet.

Meanwhile, back on the Eastern Front a more fluid situation prevailed and eventually, after a titanic three-year struggle, Germany and Austria defeated the forces of Imperial Russia and occupied the Ukraine, the Baltic States, Poland and Byelorussia. By October 1917, Russia itself had dissolved into revolution. Germany arranged to transport the communist revolutionary Lenin – in a sealed train like a plague bacillus – across Germany from Switzerland to Russia, where he was injected into Russia's body politic.

After the Treaty of Brest-Litovsk was signed in March 1917, which brokered peace between the Bolshevik government in Russia and the Central Powers, signalling the end of Russia's involvement in the war, it now became possible to move some German forces to the Western Front, though many had to be left engaged in occupation duties in the East. Germany then launched a series of major offensives in the West from March 1918, intended to beat the French and British before the Americans could deploy in force. The German offensives almost succeeded, but the Allies held, just. Reinforced by troops arriving from the United States, the Allies succeeded in pushing back the German advance. The Western Allies, with their vast empires and industrial capacity, could always resupply their battle fronts with food, munitions and men; for the Central Powers, everything was beginning to run out and morale began to collapse. By October 1918, the Austro-Hungarian Empire was close to falling apart and Germany, facing a critical situation both on the home front and the battlefront, sued for an armistice, which the Western Allies granted on 11 November 1918. The peace spelt the end of four empires: Germany, Austro-Hungary, Ottoman Turkey and Imperial Russia – the latter eventually became a communist empire.

By the time Armistice Day came, the total casualties for both sides amounted to 9,900,000 dead and over 21,000,000 wounded, out of 68,000,000 mobilised. In addition, as a direct result of the war, there were another 7,700,000 civilian deaths. The flu pandemic of 1918–20, carried by returning soldiers, would kill 50,000,000 to 100,000,000 more.

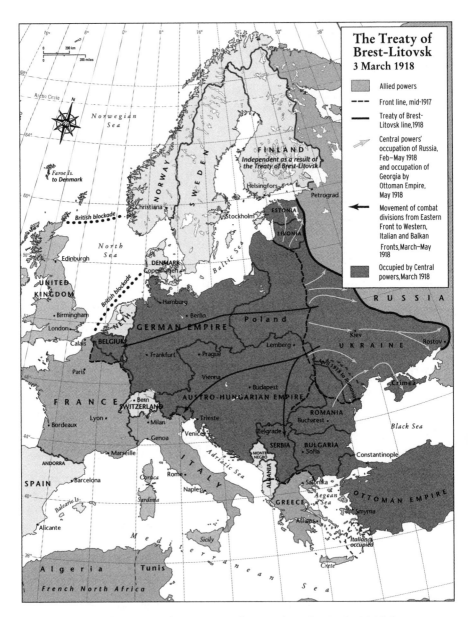

The Treaty of
Brest-Litovsk
3 March 1918

Allied powers

Front line, mid-1917

Treaty of Brest-
Litovsk line,1918

Central powers'
occupation of Russia,
Feb–May 1918
and occupation of
Georgia by
Ottoman Empire,
May 1918

Movement of combat
divisions from Eastern
Front to Western,
Italian and Balkan
Fronts,March–May
1918

Occupied by Central
powers,March 1918

*Map 55. The Treaty of Brest-Litovsk, signed on 3 March 1918 between
the new Bolshevik Government of Russia and the Central Powers,
ended Russia's participation in the First World War.*

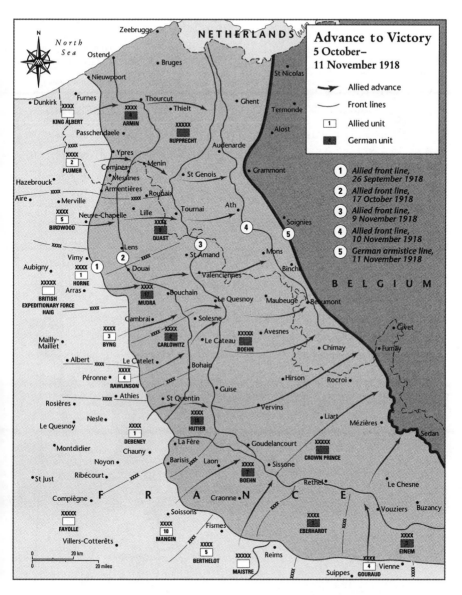

Advance to Victory
5 October–
11 November 1918

→ Allied advance

⌐ Front lines

☐ 1 Allied unit

■ German unit

① *Allied front line,*
   26 September 1918
② *Allied front line,*
   17 October 1918
③ *Allied front line,*
   9 November 1918
④ *Allied front line,*
   10 November 1918
⑤ *German armistice line,*
   11 November 1918

Map 56. By October 1918, the once-powerful German army was slowly retreating before the Allied advance, now supported by a powerful American army.

At the following Peace Conference, held at Versailles in 1919, the victorious Allies imposed territorial losses, huge financial reparations and draconian restrictions on the armed forces of the Central Powers, who would have to carry the financial cost of the war, which was calculated at $186 billion in 1919. The nature of the Treaty of Versailles would mean that ultimately this would not be 'the war to end all wars', but merely a 20-year break before the second round in the Second World War.

With the passage of time, we tend to think that these things can't happen again; I am not so sure – when I look at history, I see us making the same mistakes again and again. I have drawn the maps to prove it, and it just takes time to forget the lessons of the past.

16

THE SECOND ROUND

What we now call the Second World War began as a series of separate wars in Asia, Africa and Europe, which fused in 1941 into a global conflict. It was like no other event in all human history. The cost in destruction, especially in Europe and Asia, but also in the oceans of the world and in human life, was immense. We think some 70 million people died in the conflict, but even now new figures are coming to light, which include the effects of displacement, starvation and forced transportation on millions of souls, many of whom are still unaccounted for.

Such was the scale of this event that, of course, as a mapmaker I feel it deserves a book, or atlas, solely dedicated to explaining the almost inexplicable. I have been involved in creating four atlases on the subject and have never felt that I could come anywhere near doing justice to explaining this massive upheaval in the timeline of world history. Therefore, within the scope of this book, we are going to examine the form of this war in just four battles.

*   *   *

Following the armistice of 11 November 1918, the First World War officially ended with the Treaty of Versailles (1919). When the terms of that treaty became known, many Germans were astounded by the loss of territory and the war guilt branded on them, and, of course, they would be stuck with the cost of the war via reparations made to the victors. They would also lose their colonial possessions, which would be gained mostly by Britain, France and Japan.

As the post-war world emerged, it eventually became clear that there was deep resentment against the Versailles treaty in Germany, which was exploited by Adolf Hitler as his influence increased. There can be no doubt about the origins of the Second World War; in *Mein Kampf*, published in the mid-1920s, Hitler penned these words: 'When we speak of *Lebensraum* (living space) for the German people, we must think principally of Russia and the border states subject to her. Destiny itself seems to wish to point the way for us here.'

Meanwhile, across the world in Japan, the government was becoming increasingly bellicose. As a result of the Versailles treaty, Japan had gained former German possessions in the Pacific and in Tsingtao on the coast of China, and in 1931 Japan seized Manchuria. Japan, with its growing economy, began to develop a policy of military acquisition in order to control the raw materials and markets of the colonies belonging to Britain, France and the Netherlands. The Second Sino-Japanese War began in 1937; some academics date this as the actual beginning of the Second World War.

Back in Europe, Benito Mussolini and his fascist movement gained power in Italy from the early 1920s and began to modernise the state. Apart from making the trains run on time, Mussolini began to invest heavily in Italy's armed forces.

Italy also felt that, as an ally of the Triple Entente in the First World War, it had received little in reward for its sacrifice. In 1935, the League of Nations rejected Mussolini's aspirations for territorial aggrandisement in Africa, and this led to the Italian invasion of Abyssinia in October 1935, with the country overrun in 1936. This action distanced Italy from its erstwhile Western Allies and brought it under the political influence of an increasingly strident Germany.

After Hitler felt himself secure, his hands firmly on the levers of state and in control of the popular press, he decided to test the Western Allies by overturning the terms of the Treaty of Versailles, occupying the Rhineland, establishing the Anschluss (the union with Austria) and acquiring the Sudetenland (the German-speaking areas of independent Czechoslovakia). The war in Europe officially began on 1 September 1939 with the invasion of Poland, this time with the acquiescence of the Soviet Union following the short-lived Nazi–Soviet Pact.

The Western Allies now faced Germany, largely basing their strategy on a replay of the First World War, by constructing strong defensive lines. However, the Germans, under their radical leader, adopted a radical plan. They simply went through the Allies' defensive lines at their weakest point, the Ardennes, and broke out into the open country of northern France, cutting off the British Expeditionary Force and a large part of the French army in Belgium. The British army, with some of its allies, managed to avoid capture by retreating to Dunkirk and being rescued by an ad hoc fleet, which managed to save over 330,000 Allied soldiers.

With the surrender of France on 22 June 1940, Britain now faced Germany alone. The Luftwaffe could still command 993

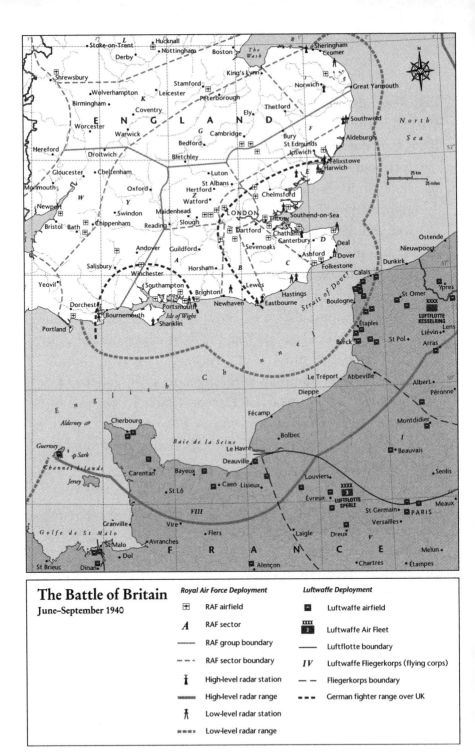

The Battle of Britain
June–September 1940

Royal Air Force Deployment		Luftwaffe Deployment	
⊞	RAF airfield	▪	Luftwaffe airfield
A	RAF sector	XXXX 3	Luftwaffe Air Fleet
—	RAF group boundary	—	Luftflotte boundary
- - -	RAF sector boundary	IV	Luftwaffe Fliegerkorps (flying corps)
Ī	High-level radar station	– –	Fliegerkorps boundary
▬▬▬	High-level radar range	- - -	German fighter range over UK
Ā	Low-level radar station		
▬▬▬	Low-level radar range		

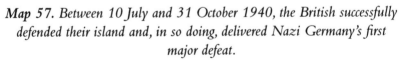

Map 57. Between 10 July and 31 October 1940, the British successfully defended their island and, in so doing, delivered Nazi Germany's first major defeat.

single-engined fighters, 375 long-range, twin-engined fighters, 1,015 bombers and 346 dive-bombers. Across the Channel the Royal Air Force had 704 single-engined fighters and a force of about 440 bombers. The main hope for the defence of Britain lay in its air defence system, based on radar, which was also supported by the Royal Observer Corps – in other words, eyeballs and binoculars – which allowed squadrons to be guided into the battle as the raid developed.

British tactics were simply to defend RAF airfields and major urban and industrial areas. The German approach was to do exactly the opposite, but they underestimated the importance played by radar. Initially, attacks were focused on Channel convoys, in order to draw British fighters into battle. However, Britain's command and control system worked too well and was almost always able to concentrate its aircraft in the right place at the right time.

On 13 August 1940, the Germans launched Operation *Adlertag* ('Eagle Day') and approximately 1,500 aircraft were sent against England. However, despite previous attacks, the British defence system was still intact and the British aircraft industry was actually outperforming that of the Germans; for instance, about 450 aircraft per month were reaching the RAF, while the Luftwaffe received just 200. And so the attacks on British air defence infrastructure changed; the main Luftwaffe target was to be London, particularly its dock areas, oil tanks and gas works. These all had nearby residential areas, which it would be impossible to avoid. Hitler authorised the raids on 5 September 1940. But the German fighter escort could only stay with bombers for 10 minutes or so, leaving them exposed to British fighter attack. At a conference of German High Command, it was understood that the RAF was still in command of the air over England and as a result daytime raids were phased out and a long-term night-bombing campaign began from 16 September. Hitler cancelled Operation Sealion (the invasion of Britain) on 13 October; the Night Blitz would continue into May 1941, by which time Hitler's gaze had turned eastwards, leaving an unconquered enemy on Germany's Atlantic flank.

Meanwhile, in September 1940, Japan had signed the Tripartite Pact with Germany and Italy, calling themselves the Axis powers, which could threaten the USSR from the east. This began the process of combining the Asian and European wars into a global war.

On the evening of 21 June 1941, the largest invasion force in history was assembled along the western border of Soviet Russia. The Axis force advancing into Russia amounted to almost 3.6 million men, with 3,600 tanks and almost 2,800 aircraft. Facing them along the western border were 2.9 million Soviet troops with between 10,000–19,000 tanks (though many were obsolete) and around 8,000 aircraft. Behind this force were another 2.6 million men across the interior of the USSR.

Stalin had received intelligence on the forthcoming attack but consistently disregarded this information, being more concerned with purging the Red Army's command structure of those people he thought politically unreliable, leaving it debilitated. The shock of the Axis attack effectively paralysed the Soviet command and control structure.

Between the end of June and October 1941, Axis forces advanced eastwards, capturing huge numbers of Soviet troops. A line was established from Leningrad to the approaches to Moscow and further south towards the city of Rostov-on-Don. Meanwhile, the Soviets had made a huge effort to move

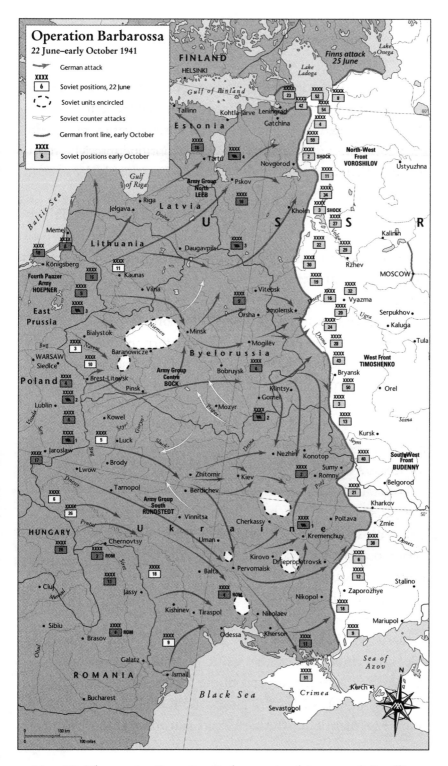

Map 58. *The massive Operation Barbarossa, involving some 3.5 million Axis troops, launched on 22 June 1941 and reached its maximum extent by November 1942.*

thousands of factories and millions of their workers east beyond the reach of the advancing enemy. To maintain morale in the civilian population as well as in the army, Stalin ordered the Revolution Day parade on 7 November. Soviet troops marched through Red Square, past the Kremlin, and then went straight to the frontline.

The winter of 1941–2 was the coldest of the 20th century. The effect on both armies was harsh to say the least but much worse for the Germans, who did not have specialist winter clothing and suffered 130,000 cases of frostbite. The German offensive wore down to a halt on 5 December 1941. Following a period of Russian reinforcement, the counter-offensive began on 5 January 1942; the German troops were unable to form a cohesive resistance and fell back to a more defensible line. Hitler gave orders to stop the retreat and defend every patch of ground, much to the anger of his generals. By the spring thaw of 1942 the counter-offensive was over; the German invaders had been defeated for the first time and the immediate threat to Moscow had been removed.

Hitler had made Moscow the final objective for this 1941 campaign. Ultimately, he failed before the defences of the city, and with that failure the idea of a quick campaign in Russia came to an end, to be replaced by the realisation in German High Command that this was going to be a long, bloody affair.

* * *

We now move across the world to Asia, where Japan had decided that the elimination of the United States Pacific Fleet was key to the successful prosecution of its planned war in Southeast Asia. The attack on Pearl Harbor was carried out on

7 December 1941, led by six aircraft carriers of the Japanese navy, with 430 aircraft aboard. They struck the American anchorage at 7.40 a.m. local time, sinking or damaging the battleships of the US Pacific Fleet, while a second wave, which arrived at 8.53, destroyed 188 American aircraft, with 2,403 Americans killed. The United States aircraft carriers, however, were all out on exercise and thus missed the attack. This was a significant factor in the subsequent prosecution of the Pacific War. In addition, the immense industrial capacity of the United States to build anew and repair damaged ships at a much faster rate than the Japanese would ever be able to achieve would, in time, affect the outcome of the now global war.

In June 1942, Admiral Yamamoto drew up plans to occupy the island of Midway in the central Pacific, and thereby draw the Americans to its defence. Yet all Yamamoto's carefully conceived plans were to no avail, thanks to the Allied ability to decipher Japanese naval codes, which enabled Admiral Nimitz, the Pacific Commander in Chief, to deploy his forces to the most advantageous positions. He sent two carrier groups: one with the USS *Hornet* and USS *Enterprise* with Admiral Spruance; the other with the recently repaired USS *Yorktown* under Rear-Admiral Fletcher. Between them they had 233 aircraft, with 127 more on Midway. Nimitz also placed a submarine screening force to the west of Midway. All this was achieved, with all ships in position, before the Japanese planned submarine screen was even deployed.

Battle began on 3 June 1942 when US aircraft, flying from Midway, spotted and attacked a Japanese transport group 700 miles west of the island. On the following day the Japanese launched 108 aircraft, targeting Midway, which was heavily attacked but remained operational. Just previous to the

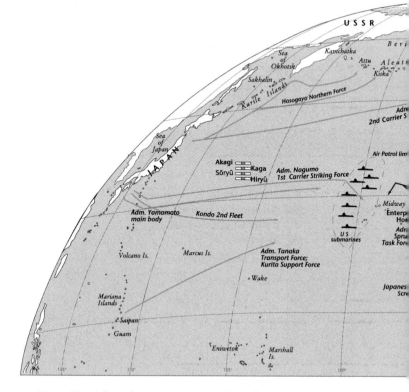

Map 59. Admiral Yamamoto's complex plan to seize Midway ultimately failed, shifting the balance of power in the Pacific in favour of the Allies.

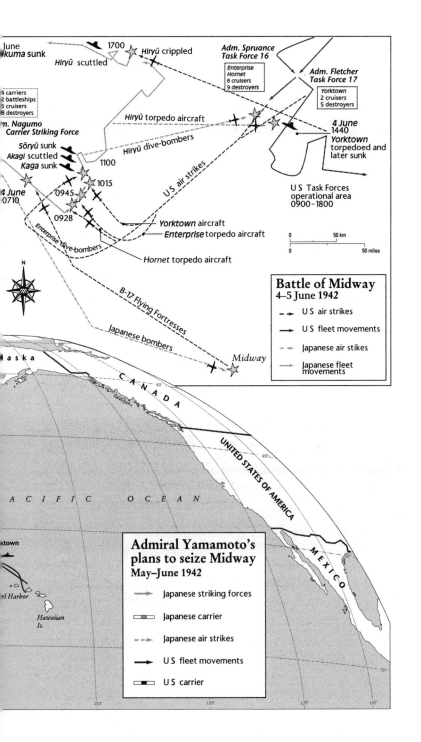

June
kuma sunk
1700
Hiryū crippled
Hiryū scuttled

Adm. Spruance Task Force 16

Enterprise
Hornet
6 cruisers
9 destroyers

Adm. Fletcher Task Force 17

Yorktown
2 cruisers
5 destroyers

4 carriers
2 battleships
5 cruisers
8 destroyers

Hiryū torpedo aircraft

m. Nagumo Carrier Striking Force

Hiryū dive-bombers

Sōryū sunk
Akagi scuttled
Kaga sunk

1100

4 June 1440
Yorktown torpedoed and later sunk

1015

4 June
0710

0945

US air strikes

US Task Forces operational area 0900–1800

0928

Enterprise dive-bombers

Yorktown aircraft
Enterprise torpedo aircraft

0	50 km
0	50 miles

Hornet torpedo aircraft

N

Battle of Midway
4–5 June 1942

- - ➤ US air strikes

⟶ US fleet movements

– – Japanese air stikes

Japanese fleet movements

B-17 Flying Fortresses

Japanese bombers

Midway

a s k a

C A N A D A

UNITED STATES OF AMERICA

45°

P A C I F I C   O C E A N

Admiral Yamamoto's plans to seize Midway
May–June 1942

Japanese striking forces

Japanese carrier

– – ➤ Japanese air strikes

⟶ US fleet movements

US carrier

town

l Harbor

Hawaiian Is.

M E X I C O

Japanese attack, US aircraft had already taken off and were en route to attack the four Japanese aircraft carriers. The location of these carriers was also passed on to the three US carriers, who immediately launched 116 aircraft in addition to those that had already left the island.

The first wave launched from Midway now attacked the Japanese carriers but failed to make a single hit on any of the ships. In the middle of this attack news came through to the Japanese carriers, under Admiral Nagumo, that an American task force had been sighted. The Japanese admiral decided to prepare a fully planned attack, while defending his ships from the American air attack that was in progress. However, before he could fully rearm and refuel his recently returned aircraft, a force of American dive-bombers appeared overhead and immediately attacked three of the Japanese carriers and the decks were soon ablaze; the *Akagi*, *Kaga* and *Sorgu* were sinking wrecks. This left the Japanese with just the one carrier operational, the *Hiryū*. Looking for revenge, the *Hiryū* immediately launched a counterattack and successfully hit the USS *Yorktown*, which was so damaged that she was put out of action. The *Yorktown* was finally torpedoed by a Japanese submarine and sank.

Meanwhile, the *Hiryū* herself came under air attack, being hit by three bombs and torpedoed twice by US submarines. She still valiantly stayed afloat and was eventually scuttled by her own crew. With the loss of four carriers, Admiral Yamamoto was forced to concede defeat and started to withdraw eastwards with the surviving ships, harassed by US dive- and torpedo-bombers. Yamamoto also realised that Japan's only plan now was a long war of attrition.

Meanwhile, back in the western hemisphere, and with America now in the war, the Axis had been thrown out of North Africa, the Allies had landed in Sicily and Italy, and now the long-planned landing in France was beginning to coalesce. The defeat of the German U-boats had allowed resources to flow across the Atlantic in order to build up the forces necessary to carry out a cross-Channel landing in France. British General Frederick E. Morgan was appointed Chief of Staff to the Supreme Allied Commander, or COSSAC. It was his job to plan what was now called 'Operation Overlord'.

Preparations were made under conditions of the greatest secrecy and a series of deception plans, called 'Operation Fortitude', were made. The first operation to take place was conducted by units of the French Resistance who attacked, in particular, communications and rail links across northern France. During the night of 5–6 June 1944, 52 locomotives were destroyed and rail lines cut in 500 places. The first airborne landing took place in the Cotentin Peninsula, just 15 minutes after midnight, by the United States 82nd and 101st Airborne Divisions, concentrating on an area just to the west of Utah Beach. Despite being badly scattered, the soldiers managed to coalesce into viable compact units and took most of their designated offensives. Meanwhile, on the eastern flank, the British airborne troops began to arrive at 16 minutes after midnight when assault gliders landed and seized Caen Canal's Pegasus Bridge, followed by the main force of the 6th Airborne Division. Once again, they were badly scattered and took some time to finally seize their objectives, which included the gun battery at Merville, which overlooked Sword Beach.

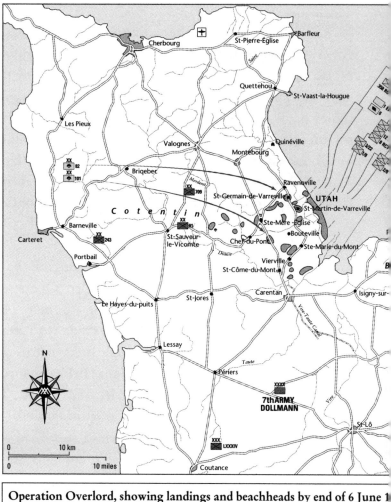

Operation Overlord, showing landings and beachheads by end of 6 June 1

Military symbols ■ German unit ⊠ Allied unit

☐ XXXX Army ☐ XXX Corps ☐ XX Division ☐ X Brigade

Abbreviations

RCT	Regimental Combat Team	**Can.**	Canadian	**Green H**	
ENG	Engineers	**SS**	Special Service	**South Ea**	
GIR	Glider Infantry Regiment	**CDO**	Commando	**N. Shore**	
RN	Royal Navy	**HQ**	Headquarters	**South La**	
RE	Royal Engineers	**Dorset**	Dorsetshire Regiment	**R. de Cha**	

Map 60. Better known as the D-Day landings, 'Operation Overlord' was the largest and most complex seaborne operation in world history.

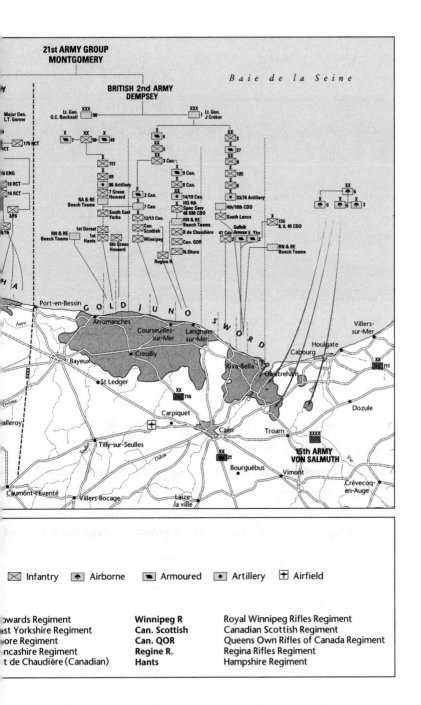

Infantry ⊠   Airborne   Armoured   Artillery •   Airfield ✛

...wards Regiment	Winnipeg R	Royal Winnipeg Rifles Regiment
...st Yorkshire Regiment	Can. Scottish	Canadian Scottish Regiment
...ore Regiment	Can. QOR	Queens Own Rifles of Canada Regiment
...ncashire Regiment	Regine R.	Regina Rifles Regiment
...t de Chaudière (Canadian)	Hants	Hampshire Regiment

The US 4th Infantry Division arrived at Utah Beach on time at 06.30, slightly adrift from their intended landing point. Nevertheless, their commander decided to begin the division's war from that very spot. They eventually met up with US airborne troops and consolidated their beachhead with just 197 casualties.

Meanwhile the US 2nd Ranger Battalion assaulted Pointe du Hoc, where a German battery commanded both Utah and Omaha beaches, climbing the 100-foot cliffs with grappling hooks and scaling ladders under German fire. They seized the position, demonstrating great initiative and bravery, only to find the German guns had been moved 600 yards further south, where they then located and destroyed them, in all suffering 135 casualties.

Omaha Beach was believed to be the most heavily defended, and therefore it was assigned to the US 1st Infantry Division and 29th Infantry Division. They were to be supported by 32 amphibious DD tanks. However, bad weather affected the timing and placing of the landings and sank 27 of the tanks. Meanwhile, the infantry landing crafts arrived, though many were blown off course and came under withering fire from German positions, which were largely undamaged. By sheer bravery US soldiers managed to form assault units from the survivors and began to fight their way inland, covered by a group of destroyers that risked running aground. The beachhead was eventually established at the cost of 2,000 casualties.

At 7.25 a.m., covered by fire from two British cruisers and support ships, the British 49th Division began landing at Gold Beach. Strong winds made the landings difficult to coordinate and it was decided to land the DD tanks closer inshore, which helped the infantry clear fortifications along the waterfront.

The only Victoria Cross awarded on D-Day was won by Company Sergeant Major Stanley Hollis for attacking enemy bunkers near Mont Fleury. Men of the Hampshire Regiment, at the cost of around 1,000 casualties, captured the village of Arromanche, which would be the site of the Mulberry harbour – a fantastic fabrication of concrete sections, which were floated across the Channel and sunk in position to create a temporary harbour, the remains of which can still be seen today.

Ten minutes later, at Juno Beach, the Canadian 3rd Division began landing. A huge bomb crater impeded the exit from the beach, so this was filled by an abandoned tank and fascines (rolls of sticks packed tightly), which formed a kind of bridge over which soldiers and vehicles could advance. Major German strongpoints, all with 75-millimetre guns and machine guns and surrounded with barbed wire and mines, were eventually overwhelmed. Other groups of defenders were well dug in and had to be outflanked before the advance could continue. Leading units of the 9th Canadian Infantry Brigade advanced to within sight of the Carpiquet airfield where they dug in for the night to await supplies. At the end of the day the casualties were 961.

On Sword Beach, the 3rd British infantry division landed at 7.30 a.m., eventually overwhelming nearby German strongpoints. Lord Lovat and his Special Service Brigade and No. 4 Commando followed up the initial landing, with the support of the Free French Commando under Philippe Kieffer, who advanced through Ouistreham, to eventually join up with the airborne soldiers at Pegasus Bridge. The 3rd Division faced the only German armoured counterattack launched on D-Day, but successfully beat it off. They suffered around 1,000 casualties.

At the end of the first day of the D-Day landings, 160,000 troops had landed in France. Germany would be caught between the advancing Allies from the west and a vengeful Red Army from the east, both growing in power as the German armies declined. The end was plain for all to see. Despite this, Germany fought on to a bloody and bitter defeat on German soil. Meanwhile in Asia the long and hard-fought campaigns over China, Southeast Asia and the Pacific would eventually be ended by the US dropping two nuclear bombs on the Japanese cities of Hiroshima and Nagasaki, heralding an end to the Second World War but introducing the dreadful possibilities of a new kind of warfare.

17

CITIES

The 21st-century city is a wonderful place. I'm sure cities have been places of wonder for as long as they have existed, but now of course the drains work and you don't catch anything when you dine out ... well, not often. The truly great cities have everything the citizen or visitor could ever wish for: theatres, libraries, galleries, museums, cinemas, beautiful parks, eateries of all kinds and a reliable police service who can help out when you get taken for a ride or mugged. (The latter happened to me just once, in New York. It was over in seconds and very efficient.)

I have a city not far from where I live now in Cheshire, which is Liverpool, once the second city of the empire, and you can tell – its architectural heritage is breathtaking. Of course, cab drivers in Liverpool have more to say than most. On a journey through an area called Toxteth, one driver said to me, 'Oh, by the way, Hitler used to live down there!' He told me that Hitler stayed for a while with his half-brother and his Irish wife. Apparently, they all used to argue and the continual racket upset the neighbours. 'I think he was trying to get into a local art school,' the driver said, 'but he packed it in and went back to Germany or Austria or some place.' This apocryphal story had certainly added a new dimension to the Führer's existence.

Liverpool's buildings across the city and, of course, the waterfront express the perfect narrative of the city's history. In most British cities, they would be labelled Jacobean, Victorian or Edwardian in style, but Liverpool needs something unique, so, because the city is so connected to Britain's activities across the oceans, I think perhaps Early Empire, High Empire and Late Empire might be more appropriate.

Some years ago, when working with my colleague, the editor Liz Wyse, I had an idea to produce a range of city histories. A perfect concept, if it worked; a series that would keep us in business for a long time and engender a feeling of wellbeing in our bank manager. The first three cities to spring to mind were, I suppose, the natural frontrunners: London, Paris and New York. I mentioned the plan to Liz, who immediately lit up – she is the most citified person I know, always off to meet friends in exotic locations worldwide. If anyone knows how to exploit a city, she does.

Of the three suggested, London sprang to life almost immediately, and with a text written by Professor Hugh Clout of University College London, we attempted the 2,000-year history of London, from its geology and first settlements to its recent past as possibly the only capital city to suffer consistent ballistic missile attack in 1944–5. One particular topic was the development of London's railways: the metropolitan railway, and in particular the Underground, known to many Londoners as the Tube, which made possible the expanding suburbs.

During the 19th century, London's population grew rapidly, not only as a national capital, but also as the centre of a worldwide empire. The Great Exhibition of 1851 attracted 6 million visitors from around the empire and the rest of the world. The subterranean system developed from 1863; the world's first, it used gas-lit wooden carriages pulled by steam locomotives between Paddington and Farringdon. The line carried 38,000 passengers on its first day of operation and was hailed as an outstanding success. The network expanded over the next 50 years, which led to the creation of probably the best-known graphic map on the planet. The famous schematic map was a revolutionary development, replacing maps such as the one featured here, which is an attempt to show the actual geography of Greater London and the underground lines. The now iconic Tube map was originally designed by Henry Charles Beck, known to most as Harry Beck, an electrical draughtsman at the London Underground signals office. You can see the influence of an electrical layout in his idea. The early plans of the system followed a strict geographical layout. This was not so easy to understand. The new map, however, caught on and has gone on to influence subway and other transport maps around the world.

The next city to spring to life was Paris, for which the text was under the care of Jean-Robert Pitte, a professor of geography at the Sorbonne. Didier Millet, the owner and founder of his own publishing company, Editions Didier Millet, took a very close interest in the conception and design of the atlas. I stepped back at this stage, looking over Liz's shoulder as she handled the somewhat complex negotiations about how our maps and illustrations might be used. The French editorial team was an enthusiastic and energetic gang and gave a significant amount of the day to their lunch break – which was more an extended meeting where the white paper tablecloths became impromptu drawing boards on which maps and layouts were

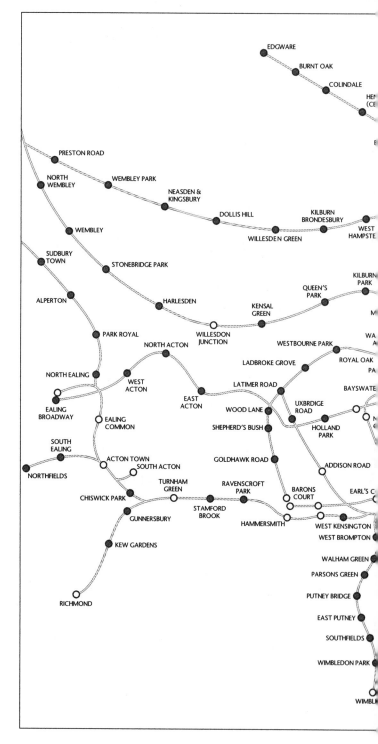

Map 61. This map pre-dates the unique diagrammatic London Underground map designed in 1931 by Harry Beck that is so familiar to us all today.

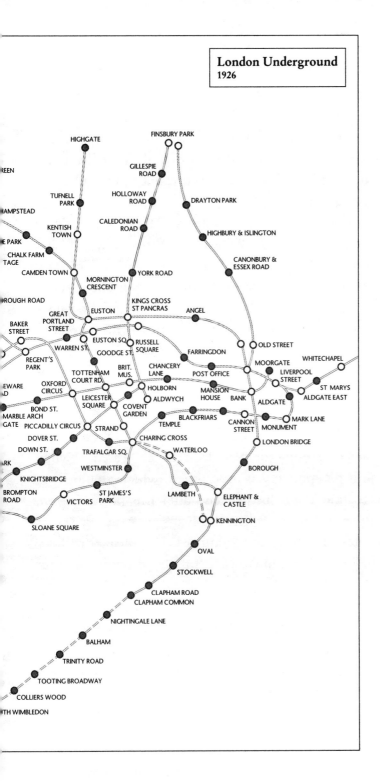

London Underground
1926

sketched. Didier, a big city man, I think had some reservations when dealing with 'provincials' like us – 'Oh, don't worry,' I assured him, 'we're on the map. We even have a direct flight between Derby and Paris.' He was astonished. Actually, it was between Paris and East Midlands Airport, but I didn't play too much on that minor point.

Paris, like London, has a long history. The first significant settlement dates from around 250 BCE when the Parisii, a Celtic tribe, settled on the banks of the River Seine. The city expanded through the ages but had experiences in the modern period unlike London, and those were revolution and occupation. In the summer of 1789, the city became the focal point of the French Revolution – an event that changed France and Europe, and left its mark on the wider world. The city was then occupied in 1814, 1871 and 1940–4, all harrowing experiences for its citizens.

We pick up Paris's story, however, on 2 September 1870 – the same day that Emperor Napoleon III abdicated – when Paris the Third Republic was proclaimed, which would serve France until June 1940. Just 17 days after this proclamation, however, Paris was tested when the Prussian army arrived at the city's gates. The city was bombarded by heavy artillery from the beginning of January and finally surrendered, after eating all the zoo animals and every rat that could be found, on 28 January. An uneasy peace existed during the Prussian occupation, but it was brief – they soon left, taking up positions outside the city.

The peace held until March when revolution broke out. Two radical soldiers of the National Guard killed two high-ranking officers of the regular army and a nervous

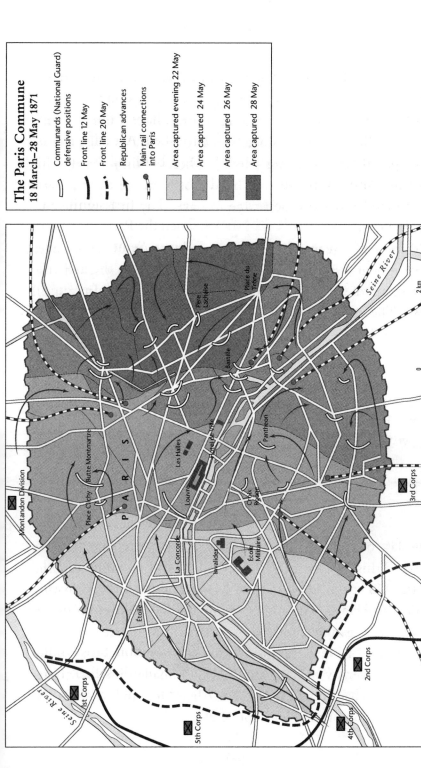

The Paris Commune
18 March–28 May 1871

- ⌒ Communards (National Guard) defensive positions
- ▌ Front line 12 May
- ┇ Front line 20 May
- ↑ Republican advances
- ●┅ Main rail connections into Paris
- Area captured evening 22 May
- Area captured 24 May
- Area captured 26 May
- Area captured 28 May

Seine River

Montandon Division

1st Corps
5th Corps
4th Corps
2nd Corps
3rd Corps

P A R I S

Place Clichy
Butte Montmartre
Étoile
La Concorde
Invalides
École Militaire
Louvre
Les Halles
Hôtel-de-Ville
Croix Rouge
Panthéon
Bastille
Père Lachaise
Place du Trône

Seine River

0   2 km
0   2 miles

Map 62. Following the Franco-Prussian War in 1871, the majority of Parisians resisted the new regime, which was forced to re-take the city.

government, its officials and units of the regular army withdrew to Versailles. Immediately, a new city council, the Paris Commune, was elected and took control on 26 March, with a radical socialist agenda. Since the early 1700s, Paris had a large poor population, scraping a living wherever they could. This insecure, shifting population lived mostly in the eastern parts of the city; now was their time to have a say in how the city might be run.

But – and it seems there is always a but – on 21 May, under government instructions, the regular army attacked the communards and by 28 May, after heavy fighting, it was all over. The commune had lost about 7,000, the army lost 837 and 6,424 were wounded. Many communards were imprisoned but later given amnesties, although only after a decade in remote prisons. Others escaped into exile. Paris recovered as always and went on to survive occupation and liberation in 1940–4. Today it is one of those special places that makes you go weak at the knees just to be there, nevermind all the treasures that exist within its bounds.

* * *

The last of the big three was New York. I am sure that Liz enjoyed this city most of all; she had spent some time teaching in the US at the University of Massachusetts after graduating from Cambridge. The author of our atlas was Eric Homberger, who was visiting professor of American literature at the University of New Hampshire and teaches American studies at the University of East Anglia, which made him easy to get at, and we had lots of help from Alice Hudson, Chief of the Map Division of the New York Public Library.

The Native Americans who occupied Manhattan Island apparently sold it to the Dutch for a few trinkets in 1626, which is where we pick up the story of New York. Or perhaps we should begin with New Amsterdam. Settlers first arrived in 1623 aboard the Dutch ship *Nieuw Nederlandt*. They were French-speaking Walloons from what is now Belgium. The Dutch had been exploring the area for several years and began settling along the Hudson River from 1614, some six years before English settlers arrived at what became the Plymouth colony in 1620. New Amsterdam grew slowly; by the mid-1630s the settlement had a flour mill, two saw mills, a shipyard, goat pens, a bakery, a church and the services of a midwife. Between 1630 and 1650, 50,000 English settlers headed for the New World with about 15,000 or so heading for New England. These numbers not only triggered a number of wars with Native Americans, but also overwhelmed Dutch and Swedish colonies in the area.

In 1664, four English frigates sailed into New York's harbour and demanded its surrender. This was agreed to and a provisional treaty was signed on 6 September 1664. The following year, in June 1665, the city was incorporated into English law as New York, being named after the Duke of York, later King James II. In 1667 at the Treaty of Breda, which ended the second Anglo-Dutch war, the Dutch confirmed ownership of Surinam in South America, while England, the losers, got to keep New York. With hindsight, that's not too bad when losing a war. The town slowly expanded and by 1790, in the early years of the new United States, it had reached a population of 33,131 (to be precise).

The old Dutch dirt road called De Heere Straat, renamed Broad Way, gave the city its direction of expansion for many

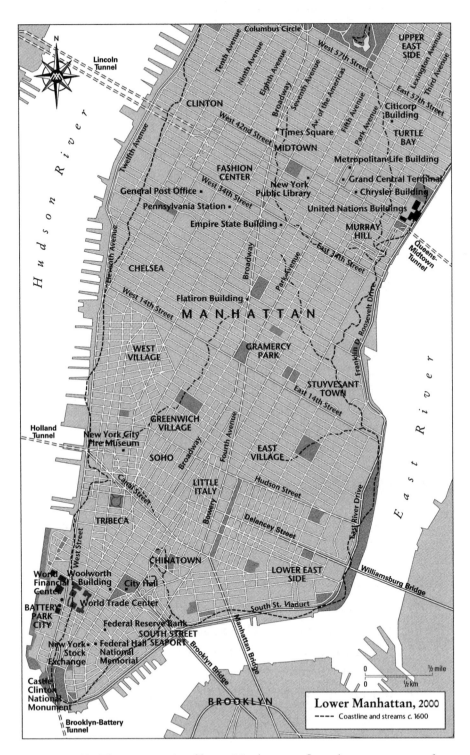

Map 63. The extreme tip of lower Manhattan reflects the street pattern of the early 1700s. From there, Broadway leads northwards, pointing to the direction of the city's expansion.

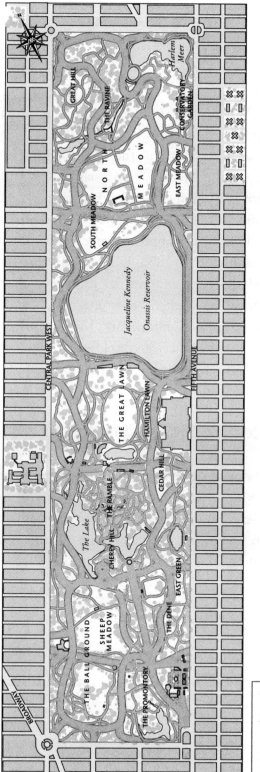

Central Park

Map 64. New York's Central Park was designed by Frederick Law Olmsted and Calvert Vaux and was opened in 1876.

decades. If you ever choose to walk Broadway and breathe in its history, wear comfortable shoes, allow time and keep your wits about you. When you get to Columbus Circle, make a left and enter Central Park.

Going back to Liverpool and the Mersey, here is a little-known fact: across the Mersey from Liverpool is Birkenhead and in that town is Birkenhead Park, the first park established at public expense in the UK. Designed by Joseph Paxton, it was opened in 1847 to the appreciation of all in the area and many from far and wide. In 1850, US journalist Frederick Law Olmsted visited the park and was so impressed that, on his return to America, he studied landscape design and became involved in the design of Central Park, the first urban park in the US, with British-American architect Calvert Vaux. It contains many features pioneered in Birkenhead, and Olmsted went on to create many civic parks across the country. Some 160 years later, Central Park is a joy to behold and after a walk up Broadway, it's a great place to eat ice cream and rest your feet.

New York is one of those cities that makes you want to skip, a bit like life with an extra gear. I spent an awful lot of my working life there, and although I don't visit so often now, I owe the city so much. It is such a creative place; I must make an effort to visit again, even if I can't skip too well these days.

18

ANOTHER VIEW OF EARTH

Ptolemy laid down detailed instructions for mathematical mapmaking in *Geography*, where he described the construction of map projections using latitude and longitude as the basic framework of mapmaking, and the tools to do this improved over the centuries. The most dramatic change in cartography was the development of aerial photography in the early 20th century. No longer was it necessary to send large numbers of surveyors and mapmakers to walk the landscape in order to prepare basic maps. Just one high-resolution camera in one airborne mission could record thousands of square miles, transforming cartography over the next 80 years or so. The process expanded when dedicated mapping satellites were launched in 1984. To this has been added computer-assisted design, followed by the development of geographic information systems (GIS), software that allows the storage and analysis of geospatial data. This allows the creation of patterns and relationships to allow cartographic designers to create maps that show comparative data, such as personal income on a state-by-state basis. Global positioning systems (GPS) add precise geographic coordinates to surface features on the earth via a network of orbiting satellites. Where would you be without your sat nav? You would have to pick up your old road atlas again – and that is a skill you should not lose. With the rise of the drone, you no longer need the expense of a large manned aircraft to collect photographic or other data; it could, and probably will, be done in a much more cost-effective way. I'm sure in a few years you will be able to check that Great-Uncle Tom has put his bins out on the right day.

The airborne activity that has been so crucial to cartographic development began with the Montgolfier brothers, Joseph-Michel and Jacques-Étienne in the 18th century. Having tinkered with the idea of hot-air balloons for quite some time, their dangerous concept was tested between 1782 and 1783 when the brothers brought together taffeta, cordage and fire. The lifting power of their balloons proved to be sufficient to take a considerable load aloft. By October 1783, in cooperation with Jean-Baptiste Réveillon, a Parisian wallpaper manufacturer, a man-carrying balloon was ready to take flight, and on or about 15 October, Jacques-Étienne was the first human being to leave the bounds of earth – even though the balloon was tethered to the ground on this flight. The first free flight was made on 21 November with two pilots (more like passengers), who must have had nerves of steel. Jean-François Pilâtre de Rozier was a chemistry teacher and would later become the first recorded fatal victim of an air accident when his balloon crashed in an attempt to cross the English Channel. The second passenger was François Laurent d'Arlandes, an officer in the French royal guard, who would later be dismissed from the army for cowardice after the Revolution – and possibly committed suicide. It seems a sad fate for such a pioneer of aviation.

King Louis XVI was not keen on risking lives on this first free flight and came up with the idea of using two condemned criminals. However, it was de Rozier who convinced the king that persons of a higher social status should be selected for this task and be remembered by posterity. So selected, they took off at 1.54 p.m. from the garden of the Château de la Muette on the edge of the Bois de Boulogne, with the king looking on; also among the onlookers was US envoy Benjamin Franklin. The flight took 25 minutes, reaching an altitude of 3,000 feet, returning to earth at Butte-aux-Cailles, which was then on the outskirts of Paris – a 5.5-mile flight. Afterwards, de Rozier and d'Arlandes celebrated with a glass or two, maybe more, of champagne. I don't know about you, but I feel the need for champagne during and after each flight; after being 3,000 feet up in a paper balloon powered by fire, I would need to swim in the stuff.

Just 10 days later, on 1 December 1783, aviation made another step forward when the first hydrogen-filled balloon took off with its designer Jacques Charles, accompanied by the inventor Nicolas-Louis Robert, at the controls. This contraption comprised a gas release valve and a bag of sand as ballast; the gas valve was opened to reduce height, while the sand was emptied to gain height. Once again, a curious Benjamin Franklin managed to get a good view from the gathering crowd.

They flew from the Jardin des Tuileries in Paris, ascending to a height of 1,800 feet, and travelled some 24 miles, landing at Nesles-la-Vallée. After landing, Charles decided on a second flight, going alone after the loss of some of the hydrogen gas, and this time he ascended to an astonishing 9,000 feet. Feeling a little odd, he released gas and gently returned to earth 2 miles away at Tour du Lay, and he never flew again. It seemed, however, that the gas balloon was a more viable option than the hot-air balloon, at least at that time.

It was the descendant of this early gas balloon that provided the platform for the first aerial photograph in history, of Paris in 1858, which, alas, has been lost. The earliest surviving photographs that we have are of Boston, Massachusetts, taken by James Wallace Black in 1860.

After developments in the photographic process, less equipment was needed to be taken aloft and the first free-flight photographic mission was carried out over Paris in 1879. All of this had obvious implications for mapmakers, who could see in the images created how their earthbound measurements and calculations compared with reality. Kites and rockets were also used to get cameras into the skies; in 1882 a British meteorologist, E. D. Archibald, was among the first to take clear, successful photographs using kites. In 1906, using a similar technique, George R. Lawrence photographed the devastation of the San Francisco earthquake.

The development of aircraft from the first powered flight in 1903 gave a new impetus to aerial photography. By the start of the First World War in 1914, most armies had some form of reconnaissance aircraft. These craft could cover a much larger area, at much greater speed, than the light cavalry who had previously performed the role. It was a French aircraft that detected the German army moving northeast of Paris in its attempt to surround the French army, according to the Schlieffen Plan, in September 1914. This allowed, in part, for the Allied counterattack on 9 September, which brought the

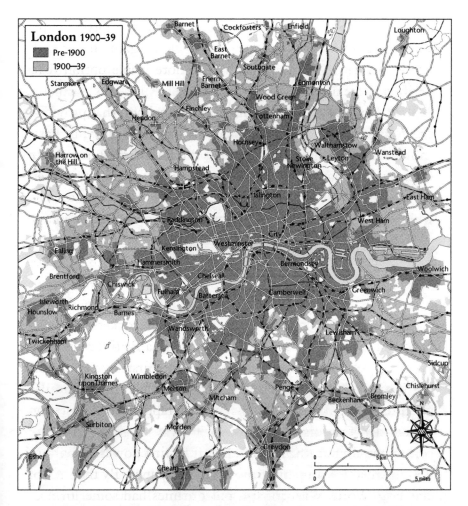

London 1900–39

- Pre-1900
- 1900—39

Barnet · Cockfosters · Enfield · Loughton
East Barnet · Southgate
Stanmore · Edgware · Mill Hill · Friern Barnet · Edmonton
Wood Green
Finchley · Tottenham
Hendon
Hornsey · Walthamstow · Wanstead
Harrow on the Hill · Stoke Newington · Leyton
Hampstead
Islington · East Ham
Paddington · West Ham
Ealing · Kensington · Westminster · City
Hammersmith · Bermondsey · Woolwich
Brentford · Chelsea
Chiswick · Fulham · Greenwich
Isleworth · Battersea · Camberwell
Hounslow · Richmond · Barnes · Wandsworth
Twickenham · Lewisham
Sidcup
Kingston upon Thames · Wimbledon · Merton · Penge · Chislehurst
Mitcham · Beckenham · Bromley
Surbiton · Morden
Esher · Croydon
Cheam
0 5 km
0 5 miles

Map 65. After 1900, the metropolis of London expanded rapidly. During this period, airborne surveys facilitated an increasingly accurate representation of the urban environment.

German advance to an abrupt halt and became known as 'the Miracle of the Marne'.

During the war, the quality of airborne cameras improved in size and focal quality. By 1918 both the Allies and Germany deployed sufficient aircraft on the Western Front to photograph its entire length twice a day. By the war's end, just over half a million images were made of this long front line, which reached from the Belgian coast to the Swiss border. This area remains the most photographed section of the earth's surface, which in turn produced the most accurate military maps attempted to date.

*　　*　　*

To some extent we almost all have that aerial déjà vu moment when we travel by air and look down. Coming into Heathrow, London, you see the familiar shape of Docklands, the curves of the River Thames, Buckingham Palace and Parliament – all just as the map illustrates. I had a moment like that on a flight out of La Guardia, New York. It was a small 16-seater commuter aircraft (the sort of aircraft I just hate – they are just not designed for someone who is six and a half feet tall). After crawling aboard with moans and groans, I managed to jam myself into a seat designed for a child. Meanwhile, the regulars all filed on without mishap. The pilot, Bill, seemed to know most of the passengers; he shook hands and said howdy to many of them. We were heading to Ithaca in upstate New York. Briefly after take-off, we circled the city and Bill announced that we had been diverted out over Long Island, where we would have to hold for a while. After fresh groans from the passengers, we settled down. This interlude, however, gave great views over the area of General Washington's campaigns of 1776 at the western end of Long Island and on Manhattan, culminating at the battle of Harlem Heights on 16 September, an American victory. There it all was, in fantastic detail.

At last we were clear to head northwest to Ithaca and Cornell University. I was going to see Professor Carl Sagan, the astronomer and astrophysicist, whose skills enabled him to explain complex subjects to a wide audience. His mapping requirements were much more contemporary. I was visiting on behalf of his publishers, who were developing a new title, provisionally called *Nucleus*. He was just finishing up with a couple of students when I arrived and, after introductions and a brief chat, he suggested we go over to his place, a short drive away. I followed in my rental car and pulled up, parking along the edge of a wooded lane. He was standing by a gate in the hedge and indicated I should follow him down a short path. In a clearing stood what seemed like a replica of an Egyptian temple – not huge, but still imposing. We entered the building through an ornate door, passed through the lobby and into a large room that had books covering two walls and a scattering of comfortable chairs. We eventually got on to the problems he foresaw in the forthcoming book and the graphics and maps he needed, and I went through the sketch ideas I had brought along with me. He looked through these, stopping occasionally to make a sort of 'mmm' sound, and after pondering for what seemed like an age, he said that they certainly went a long way towards solving the visual problems.

The far side of the room overlooked a steep chasm, with a waterfall at one end that poured into a dark space below – such a dramatic setting for his temple. This surreal, timeless building seemed an ideal place to ponder the future of humans and their possible discoveries in the infinity of space; the final frontier, you might say. I could sense my time allowance was used up, so, thanking the professor, I left.

There is a famous photograph of earth called *Earthrise*, taken by Bill Anders from the Apollo 8 spacecraft in 1968. The photograph shows planet earth floating in the vastness of space, with the surface of the moon in the foreground. The globe that ancient cartographers could only imagine is a reality of continents and oceans, the shapes of which are now so familiar to us. It looks so delicate and vulnerable out in the cold vastness of space. I like to think that may have been the genesis of Professor Sagan's *Pale Blue Dot: A Vision of the Human Future in Space*, and maybe some ideas we discussed made it into that book.

Events as they unfold and become history provide endless challenges to the mapmaker to represent the human story. Over the last few decades, spacecraft have been sent on missions around our solar system to map our nearby planets, and perhaps in the future artificial intelligence can create continuously updated, statistically based maps of world events – for instance, world migrations, population growth and climate change. Data will be collected and maps updated without further human intervention. But I'd like to think that cartographers like myself will continue to reveal and interpret the essential nature of world history and represent it in a way that most of us can understand. As has already been said, there exists within all of us what psychologists call 'cognitive mapping' – the ability to process spatial data. Maps reflect our ability to both understand the world around us and relate to it, and mapping has a unique role in understanding our past and pointing the way to our future.

ACKNOWLEDGEMENTS

This story was a long time in the making and therefore there are a lot of people to thank, starting with my parents. My mother always stressed neatness of turnout and a timely approach to my schooling and the workplace. Alas, in this I did not always succeed, but when I failed at least I felt guilty. My father always encouraged my imagination and passed on the urge to find out what was over the next hill, and generously indulged my drawing skills with an endless supply of paper and pencils.

My schooldays were, from an academic point of view, a little bleak, apart from the help of a rather lofty headmaster, Mr Beasley, who always called me 'Watson!' and allowed me a little variation in the curriculum the officials laid down. I should explain that my schoolmates – good lads, one and all – were mostly sons of the soil, from farms around the district, who would tolerate learning to read, write and add, but as for the rest, it was a chance for a chat or a look out of the window. That was until the arrival of our new history teacher, Miss Redmond, a raven-haired beauty fresh from teacher training school whose film-star looks held my classmates spellbound, while I managed in the unusual silence to ask a few thousand questions over the years. I received lucid and considered replies, and at the end of term I got 94 per cent in history, becoming teacher's pet. So thank you, Mr Beasley and Miss Redmond.

I met Candida Geddes (as she was then), my first editor, when she was working for Times Books, and so my life of bringing together history and the graphic arts began. Barry Winkleman, MD of the same company, became a regular conspirator in the creation of new atlases. Influences became numerous as the number of titles I created grew. I will mention a few: the late Jonathan Riley-Smith, University of London; the late John Pimlott, Royal Military Academy Sandhurst; the late Lord Alan Bullock, St Catherine's College, Oxford; Mark Carnes, Columbia University, New York; Malise Ruthven, late of Cambridge University and the BBC and an independent author of renown. All of them could deliver a concise 10-minute telephone lecture in answer to an enquiry; what more could a boy from the backwoods ask for?

I'd like to thank my friends, Jeanne Radford – whom I have known since childhood, and who has worked with me for 40 or so years and is never short of an opinion – and Elizabeth Wyse, who read the manuscript in detail and helped turn it into a more readable script. My son Alexander took time out of his regular daily tasks to draft the maps in this work.

Oliver Malcolm of HarperCollins thought my story worthy of putting into print, brave man, and Zoë Berville, editor at HarperCollins, gives life to the saying 'Patience is a virtue'.

Finally, my thanks to my dear wife Heather, who maintained an atmosphere of calm progress; how she did that beats the heck out of me.

All the opinions and ideas expressed in the work are my own and I accept all responsibility.